Q 183.3 A1 S3515 2007 c.2

110630783

SCIENCE *and*
the UNIVERSITY

Science and Technology in Society

DANIEL LEE KLEINMAN
JO HANDELSMAN

Series Editors

SCIENCE *and*
the UNIVERSITY

Edited by

PAULA E. STEPHAN

and

RONALD G. EHRENBERG

THE UNIVERSITY OF WISCONSIN PRESS

The University of Wisconsin Press
1930 Monroe Street, 3rd Floor
Madison, Wisconsin 53711-2059

www.wisc.edu/wisconsinpress/

3 Henrietta Street
London WC2E 8LU, England

Copyright © 2007
The Board of Regents of the University of Wisconsin System
All rights reserved

5   4   3   2   1

Printed in the United States of America

Library of Congress Cataloging-in-Publication Data
Science and the university / edited by Paula E. Stephan
and Ronald G. Ehrenberg.
p. cm. — (Science and technology in society)
Includes bibliographical references and index.
ISBN 0-299-22480-5 (cloth : alk. paper)
1. Science—Study and teaching (Higher)—United States.
2. Research—United States—Finance. 3. Graduate students in science—
United States. 4. Science students—United States. 5. Universities and colleges—
United States. I. Stephan, Paula E. II. Ehrenberg, Ronald G. III. Series.
Q183.3.A1S3515   2007
507'.073—dc22   2007011939

To our spouses,

BILL *and* RANDY

Contents

Preface

Scientific research has come to dominate many American university campuses. The growing importance of science at the turn of the twenty-first century is due to exciting breakthroughs in biology, information technology, and advanced materials that have promise of tremendously improving human well-being. The growing importance of science has been accompanied by a growing flow of funds to universities from government, corporations, and other donors to help support the universities' increased scientific research. Every major university wants to achieve the prestige that comes from being a leader in scientific research, which in turn helps it attract the best undergraduate and graduate students, top faculty, increased external support for its faculty members' research, and increased annual giving. As a result, universities have increasingly poured their own institutional resources into the support of their scientific research enterprise.

This growing importance of science to American research universities has led to a number of important policy issues that should be of concern to universities and to public policymakers. These include how universities fund their scientific research, how the federal government allocates funds for research to universities, how corporations interact with universities to fund academic research and then to commercialize faculty members' research findings, who currently conducts scientific research at American universities, and who will conduct this research in the future. Many of these policy issues are best viewed in an international context, given the important role that the foreign-born play in U.S. science—both as graduate students and as scholars—and the increased competition that U.S. science is experiencing from abroad.

Science and the University brings together scholars spanning a wide variety of disciplines—including biology, economics, law, management, plant breeding, and public policy—to address these issues. Preliminary versions of their chapters were originally presented at a conference sponsored by the Cornell Higher Education Research Institute (CHERI) in May 2003, and the volume has benefited from the comments the authors received at that meeting.

Financial support for CHERI is provided by the Andrew W. Mellon Foundation, the Atlantic Philanthropies (USA) Inc., and the TIAA-CREF Institute, and we are grateful to them for their support. Publication of this volume was facilitated by funding from the Department of Economics, Georgia State University, and from a gift by William Frey to Cornell University, and we are also most appreciative of this support.

PAULA E. STEPHAN AND RONALD G. EHRENBERG
Summer 2007

SCIENCE *and* *the* UNIVERSITY

Introduction

Science and the University

PAULA E. STEPHAN AND RONALD G. EHRENBERG

THE CONTEXT

Science and engineering (S&E) research contributes significantly to the over-all quality of life in a society. For example, research creates new opportunities that lead to economic growth. In the biomedical area, research has done much to extend life expectancy. Research also helps to address humanity's quest for basic understanding; the mission to Mars in 2004 that discovered the presence of water on that planet is but one recent example of this quest.

Three sectors of the economy—industry, academe, and "other" (which includes federal and state government and nonprofit research labs)—conduct S&E research. *Science and the University* focuses on the academic sector and examines the current status of the science and engineering enterprise in this sector, as well as issues that affect the future course of S&E research in academia.

The majority of university research occurs in laboratories that are directed by university faculty members and staffed by graduate students and postdoctoral fellows. Responsibility for funding these laboratories is generally the responsibility of the faculty member. A record of successful funding and publications that derive from the research that the funding permits is often a necessary condition for obtaining promotion and tenure at a research university. Since World War II, the lion's share of the funding has come from the federal sector, although the amount from the federal government has been subject to fits and starts, both by field and over time (Dickson 1988). Recently, we have seen this in the doubling of the National Institutes of Health budget at a time when federal funds for research in the physical sciences have waned. Partly in response to the increased expense of doing science, and partly in response to the fits and starts of federal funding, the share of research funding provided by universities out of their own internal resources has increased over time. The share provided by industry has also been on the rise, partly in response to the same factors.[1]

Funding opportunities and politics, at both state and federal levels, have considerable influence on the research foci of universities. Historically, many states directed university research to targeted subjects: Wisconsin focused on dairy products, Iowa on corn, Colorado and other Western states on mining, North Carolina and Kentucky on tobacco, Illinois and Indiana on railroad technologies, and Oklahoma and Texas on oil exploration and refining (Goldin and Katz 1998, 1999; Rosenberg and Nelson 1994). Beginning with World War II, defense-related funding altered the focus of university research and led to the expansion of several universities, including the Massachusetts Institute of Technology and the California Institute of Technology. Other universities learned from the experiences of their sister institutions and used postwar defense contracts to propel themselves into the all-star league. Stanford was an early postwar example of this; more recently the Georgia Institute of Technology and Carnegie Mellon have benefited from defense-related research (Leslie 1993, 12). The tremendous growth in biomedical research funds has also contributed to the growth of universities with a heavy focus on medical-related research, such as the University of California–San Francisco and Johns Hopkins University.[2]

In terms of a metric of performance, in 2004 the United States spent approximately $312 billion on research and development (R&D); of this total, approximately 14 percent was performed in universities, 70 percent in industry, and the remaining 16 percent in the "other" sector (National Science Board 2006, vol. 1, fig. 4.2). The relatively small percentage of R&D performed in academe belies the important role that universities play in the performance of basic, as opposed to applied, R&D. When basic research is separated out, one finds that universities are responsible for approximately 54 percent of all basic research in the United States ($31.8 billion out of a total of $58.4 billion in 2004), while industry is responsible for only 16 percent of all basic research (National Science Board 2006, vol. 1, table 4.1).

The terms "basic" and "applied," while useful for the classification schemes of government statistical agencies, oversimplify the research process and reasons for doing research. Donald Stokes (1997) notes that much of today's research is both "use inspired" and inspired by a quest for fundamental understanding. In honor of Louis Pasteur, Stokes classifies such research as falling into "Pasteur's Quadrant." Stokes argues that scientists increasingly work in Pasteur's Quadrant, in part because of the scientific opportunities that have become available in recent years in such areas as biotechnology and, to extend his argument, nanotechnology. Stokes contrasts this to research that falls in "Bohr's Quadrant"—research that is motivated exclusively for fundamental understanding—and research in "Edison's Quadrant"—research inspired exclusively by use.

It is also an oversimplification to assume that research occurring in the university sector is distinct from that occurring elsewhere. The research boundaries between universities and other sectors are porous and, as we shall see, are becoming increasingly so. Michael Gibbons and his coauthors (1994) see this as one characteristic of what they call Mode 2, a new mode of knowledge production, which they argue is distinct from Mode 1, where research is done within the university, within disciplinary boundaries, and is homogeneous and hierarchical. By contrast, "the new mode operates within a context of application in that problems are not set within a disciplinary framework. It is transdisciplinary rather than mono- or multi-disciplinary. It is carried out in non-hierarchical, heterogeneously organized forms which are essentially transient. It is not being institutionalized primarily within universities structures" (vii).

While there is considerable debate over some of these claims (for example, the "newness" of Mode 2 [Pavitt 2000]), it is clear that university researchers work with researchers outside their own disciplines. It is also clear that university researchers are heavily influenced by the research and technological opportunities that occur outside the academy and frequently work with scientists and engineers located outside the university. Moreover, this cross-sectoral work often enhances the research activity of academic scientists and engineers. Lynn Zucker and Michael Darby, with coauthors (1998a, 1998b), for example, find that the productivity of academic scientists is enhanced when they work with scientists in biotechnology companies; Mansfield (1995) found that academic researchers with ties to firms report that their academic research problems are frequently or predominately developed out of their industrial consulting and that the consulting also influences the nature of work they propose for government-funded research.

It is also important to recognize that research is heavily influenced by technological advances and opportunities. While many of these technological breakthroughs are developed outside the university sector, some important technological advances have occurred within the university. Genetic research has been profoundly affected, for example, by the ability to sequence genes, a technology developed by LeRoy Hood while he was at the

California Institute of Technology. The MRI, which grew out of research in physics, has had a profound impact on research in a number of fields. More generally, as Rosenberg has so eloquently argued (2004), causality runs not only from science to technology but from technology to science.

There are several ways to place the importance of university research in perspective. One way is to use metrics of performance and outcome measures. A second focuses on spillovers between the university sector and the larger society, examining, for example, the role that university research plays in creating value outside the university. Still a third examines the role that universities play in educating future scientists.

Publication counts provide an outcome metric. While industry may perform 70 percent of all R&D in the United States, academe publishes almost 75 percent of all articles (74.1 percent in 2003) and industry publishes only about 7 percent (National Science Board 2006, vol. 1, table 5.19 and fig. 5.51). Moreover, the academic share of publications has grown slightly over time, and the industrial share has declined. Another outcome metric is the patents received for new technologies; universities play an important and growing role here. In 2003, 3,259 patents were awarded to academic universities; twenty years earlier only 437 patents were awarded to academic institutions (National Science Board 2006, vol. 2, table 5.68).

The external research funding level of a university, the publications of its faculty members, and the number of patents that it generates in a year are often used by the university to see how it stacks up relative to its competitors.[3] These measures, however, understate the important role that universities play in contributing to economic growth and quality of life. Although the wider impact of university research is indisputable, the process by which knowledge spills over from the university sector to other sectors is often long and circuitous. There are, however, several indicators of this spillover process.

One involves an examination of patent applications, the vast majority of which come from industry and which require investors to reference journal articles related to the patent. When citations to these journal articles are analyzed, it turns out that 61 percent of all articles cited in patents are written by the academic sector (National Science Board 2004, vol. 2, table 5.53). A number of researchers have also shown that the amount of innovative activity occurring in the private sector in a geographic area is related to the amount of research being performed in nearby universities.[4] Other research examines how the performance of firms relates to ties with university researchers. Zucker and Darby (1998b), for example, find that biotech firms in California doing joint research with highly productive university faculty are more successful on a variety of measures than are similar firms that are not doing such joint research.

Perhaps the most convincing work to establish the relationship between research and economic growth was done by James Adams (1990). Adams found that productivity growth in twenty industries was positively related to articles published in scientific journals in associated academic fields. Moreover, he also found very long lags between the publication of research and the impact of that research on the economy, with lags on the order of twenty to thirty years. Even in fields such as the life sciences, where time lags are becoming much smaller, there is an extended period of time before the results of the research have a significant impact on economic growth. The implication of this is that if a decline in research funding and scientific output occurs today, many years may elapse before the impact is felt on the economy.

Universities also contribute to the science and engineering enterprise in the United States by educating the future S&E workforce. While some of the students that United States universities educate in science and engineering will eventually be employed at colleges and universities, by far the largest group will head to the private sector. Others will work in the government or for the nonprofit sector. The placement of former students is another means by which knowledge is transmitted between the university sector and other sectors of the economy. The knowledge embodied in recent graduates can be especially important to industrial R&D laboratories because it facilitates the transmission of tacit knowledge, which by its very nature cannot be codified in journal articles but instead is best passed on through learning by doing.

WHY THIS VOLUME

A fundamental theme of *Science and the University* is that tremendous changes are occurring in the practice and funding of science and engineering at American universities and that these changes have important implications for universities and for the future of the U.S. science and engineering enterprise. Change can be seen in at least eight specific areas:

1. The composition and number of Ph.D. students;
2. The composition of the scientific workforce;
3. The cost of doing science at universities;

4. The way in which university research is funded;
5. The intellectual property rights of universities;
6. The measured productivity of U.S. scientists;
7. The way research is conducted, including the staffing of labs;
8. The international context of U.S. science.

Science and the University documents these changes and discusses the relationships among these changes. A case in point is the relationship between the increased cost of doing science at universities and increased class size, which in turn may discourage domestic undergraduates from following careers in science and engineering.

Some of the research presented in this volume uses data from the National Science Foundation that has heretofore not been readily available for researchers to analyze at the micro level. Richard Freeman, Emily Jin, and Chia-Yu Shen, for example, use micro-level data from the Survey of Earned Doctorates to look at changing enrollment patterns in graduate school; Grant Black and Paula Stephan use data from the same survey to examine the propensity of newly minted Ph.D.s on temporary visas to return to their home country; George Borjas uses micro records from the Survey of Doctorate Recipients to examine the extent to which foreign-born graduate students are a substitute for domestic graduate students; and Stephan and Sharon Levin use data from several NSF-administered surveys to analyze the extent to which universities are substituting foreign-born faculty for U.S.-born faculty.

Science and the University also investigates the consequences of the changes noted above for U.S. science and engineering, as well as for those doing science and engineering research at U.S. universities. By way of preview, Ronald Ehrenberg, Michael Rizzo, and George Jakubson inquire as to who pays for the increasing costs of doing university science, especially those related to start-up costs. John de Figueiredo and Brian Silverman's chapter asks whether the increasing practice of earmarking federal funds for specific universities, as opposed to awarding federal research funds through competitive processes, leads to lower-quality research. Borjas examines whether the dramatic increase in enrollments of foreign graduate students comes at the expense of American students. Stephan and Levin ask whether there is any evidence that noncitizens are being substituted for citizens for jobs in academia. Marie Thursby and Jerry Thursby's chapter queries whether the Bayh-Dole Act, by encouraging faculty to patent, has led to a decline in basic research at universities.

An Overview

Much of the change that is occurring in university science needs to be thought of in an international context. Diana Hicks, for example, argues that the decrease in the number of U.S.-authored articles published in major scientific journals is due to increased competition for journal space from scientists and engineers in other countries. She speculates that this increase has been brought about by increases in funding for science and engineering research in other countries, as well as by changes in the incentive structures in these countries. A Chinese institute, for example, provides a $600 bonus to any scientist who publishes in two prestigious journals (*Science* and *Nature*). South Korea and China have been particularly aggressive in expanding their support for research and development. Other nations, such as some in Latin America, have made efforts to expand their support. The end result, according to Hicks, is that "American universities no longer stand alone at the scientific frontier." Or, to put it in the words of a *New York Times* headline, the "U.S. Is Losing Its Dominance in the Sciences" (Broad 2004).

The changing composition of our Ph.D. student population must also be viewed in an international context. The number of science and engineering Ph.D.s granted by U.S. universities grew over the last twenty years. This increase was due in large part to a dramatic increase in the number of non-native Ph.D. candidates studying in U.S. universities, as well as to a dramatic increase in the number of women receiving Ph.D.s in certain fields. Black and Stephan document the significant increase in the number of noncitizen students in American Ph.D. programs and how the propensity of noncitizen Ph.D.s to remain in the United States after receiving their degrees has changed over time. Their findings relate to Hick's work: students from countries that are now investing more in science and engineering are more likely to return to their home countries after receiving their Ph.D.s than are students from countries that have historically invested little in science and engineering. Black and Stephan also find that new Ph.D.s from foreign countries are more likely to stay in the United States after graduation if they study at top-rated Ph.D. programs. Borjas suggests that while this increase in foreign-born students has little effect on the probability that a typical native-born student is enrolled in a Ph.D. program, the increase in foreign-born Ph.D.s has adverse effects on the enrollment of white male citizens in Ph.D. programs, and this effect is strongest at the most elite universities.

U.S. universities also face increasing competition for graduate students. Beginning in the mid-1990s, the European Union (EU) eclipsed the United States in terms of the absolute number of Ph.D.s awarded in science and engineering (NSF), and both the EU and the United States are now being challenged by the increased number of Ph.D. programs in Asian countries—especially China and Korea. The challenge is especially intense for the United States, which has seen applications and (to a lesser extent) admissions decline since 9/11 (National Academy of Sciences 2005b).

Stephan and Levin take a different look at the international question, weighing the costs and benefits related to the presence of foreign-born scientists and engineers in the United States, regardless of whether their degrees were earned at U.S. universities or not. Their research suggests that, at least in the recent past, exceptional contributions to U.S. science have been disproportionately made by foreign-born and foreign-educated individuals, indicating that these individuals contribute disproportionately to U.S. scientific output. But they caution that this may be in the process of changing. In terms of costs, they present evidence that foreign-born Ph.D.s who are employed in the United States are being substituted for American citizens for jobs in academe. However, this substitution comes largely in less prestigious temporary positions, including postdoctoral appointments.

Other changes in university science and engineering are specific to the United States. High on the list is how the passage of the 1980 Bayh-Dole Act has affected the practice of scientific research at U.S. universities. While not all of the increase in university patenting activity can be attributed to the passage of the Bayh-Dole Act (Mowery et al. 2001), Bayh-Dole has clearly led universities to adopt an aggressive technology-transfer program. Does the ability of universities to seek intellectual property protection encourage faculty to alter their research agendas and to focus more on generating patents than publishing research findings? Does it encourage faculty to substitute applied research for basic research? Risa Lieberwitz, a lawyer by training, argues in her chapter that the answer is yes, citing a number of surveys that address altered research practices and procedures at universities. Thursby and Thursby investigate this question using data for more than 3,000 faculty members employed at six major research universities during the 1983–96 period. Their research is noteworthy in that they are the first to have longitudinal data, at the faculty level, to study the relationship between patenting and publishing. They find a dramatic increase in the propensity of faculty members to "disclose" their research findings (the first stage in the university patenting process) during this period. However, the proportion of these faculty members' publications appearing in basic scientific journals has remained relatively constant during the 1983–96 period. Thus, in contrast to Lieberwitz's concerns, their findings suggest that the composition of research has not been altered as a result of increased financial incentives for faculty to engage in the commercialization of their research.

Lieberwitz is also concerned with the increased propensity of universities to receive funding from industry and the implications of such funding for the free exchange of ideas, a question that has generated widespread inquiry among scholars in recent times.[5] Ronnie Coffman, William Lesser, and Susan McCouch address what these changes mean for the future of one area of science, plant breeding, arguing that if present trends continue in the patenting of genes, "two or three companies will have a major influence on the global food system."

Scientific research conducted at universities has also become more and more expensive. What is often unclear to many is that universities themselves are increasingly bearing the burden of these increased costs, some of which stem from the large start-up packages that universities now offer to attract new faculty members in science and engineering. Ehrenberg, Rizzo, and Jakubson document these increasing costs, basing their estimates on a survey that they conducted during the summer of 2002, and they investigate who within the university bears the costs of these internal research expenditures. Their research suggests that although students may benefit from being in close proximity to great researchers, they pay for this proximity—in terms of higher student/faculty ratios and, in private institutions, somewhat higher tuition levels. Their work helps explain the changing "mix" of faculty. In other work, Ehrenberg and Zhang (2005) have shown that the high price of start-up packages affects the willingness of universities to hire at the tenure rank; it also encourages universities to replace full professors with assistant professors. Their findings may also help to explain the growing efforts of universities to commercialize faculty research. Forced to bear more and more of the costs of research, universities are looking for sources of revenue to offset these expenditures.

External research funding at universities comes largely from the federal government. Funding is often awarded on the basis of peer review—especially funding coming from the National Science Foundation and the National Institutes of Health. De Figueiredo and Silverman document how the mix of federal funding is changing and show that earmarking of grants by Congress is becoming increasingly prevalent. The recipient universities are not bashful; the authors suggest that lobbying, especially if the university has a senator or congressman who represents it on the relevant appropriations committee, greatly increases the probability of receiving an earmark. But universities pay a price. Research suggests that earmarking results in lower-quality research, as measured by citation counts.

Another theme of *Science and the University* is the changing nature of the way scientific research is conducted at universities. Laboratories are becoming bigger and scientific equipment is becoming more expensive. Increasingly, and consistent with Michael Gibbons et al.'s notion of Mode 2, discoveries are being made by cross-disciplinary research teams. These trends are discussed in the chapters by Coffman, Lesser, and McCouch and Susan Gerbi and Howard Garrison. These changes have broad implications for the university, including how laboratories should be staffed. Gerbi and Garrison argue that continued heavy reliance on graduate students and post-doctoral fellows creates problems. They suggest that universities should, instead, shift to a system in which the staffing of academic laboratories increasingly makes use of permanent research scientist positions and that these positions carry salary and benefit levels that will make long-term employment in them be seen as an "honorable career."

A final theme in the volume is how the composition of U.S.-born graduate students has changed over time in terms of gender, undergraduate institutional selectivity, and doctoral program selectivity. Jeffrey Groen and Michael Rizzo demonstrate the richness of insights that can be gained in analyzing the level of graduate school attendance of a cohort of college graduates by decomposing the cohort's graduate school attendance into its propensity to attend graduate school and the number of students in the cohort "at risk" of attending graduate school. They show, for example, that the sharp decline in graduate school attendance among American men that occurred in the early 1970s related to a decline in the group's propensity to attend graduate school, most likely occasioned by the end of Vietnam-era draft deferments for graduate school attendance. In contrast, the significant increase in Ph.D.s received by American women has occurred largely because of the growing number of American women receiving undergraduate degrees in the United States, not because of a substantial increase in the propensity of female college graduates to enroll in graduate school.

Groen and Rizzo also show that changes in the propensity of graduates of top liberal arts colleges and research universities to go on for Ph.D.s tracked the changes in the propensities of college graduates more generally to go on for Ph.D.s during the 1970s and 1980s. However, in recent years, these propensities of students from top undergraduate institutions to go on for Ph.D. study have increased relative to those from other colleges and universities. Freeman, Jin, and Shen document the widening range of institutions in the United States that send students on to Ph.D. study; the share of American students receiving Ph.D.s whose undergraduate degree comes from top undergraduate colleges and universities is declining. Similarly, the share of Ph.D.s being produced by top graduate programs is declining. The numbers of Ph.D.s being produced by universities such as Harvard and the University of California, Berkeley, have remained roughly constant over time, while Ph.D. programs at lesser-ranked universities, especially those in newly created programs, have increased their number of Ph.D.s over time.

STATISTICAL METHODS

A few words on statistical methods are in order before turning to the contributed chapters. All of the chapters in *Science and the University* have been written in a style that should be accessible to readers without any technical background in statistics. However, whenever new empirical research is summarized, as is done in a number of the chapters, invariably some statistical "jargon" and technical details creep into the chapter. While readers can skip these details, because all of the authors summarize their findings in a nontechnical fashion, what follows is an effort to provide interested readers with some guidance about the methods used in some of the chapters.

The chapters by Ehrenberg, Rizzo, and Jakubson (chapter 1), Thursby and Thursby (chapter 4), Black and Stephan (chapter 6), and Borjas (chapter 7) all use multivariate regression analyses. Such analyses are necessary whenever it is hypothesized that a number of different explanatory variables all simultaneously influence an outcome of interest and when the researcher is interested in learning what the effect of a specific explanatory variable is on the outcome. Multivariate regression analyses allow the researcher to obtain an estimate of the effect of a one-unit change in each explanatory variable on the outcome, holding constant all of the other variables that are postulated to affect the outcome.

The regression coefficients that are reported in these chapters are the authors' estimates of these effects. They are not known with certainty, and in addition to providing mean estimates for each variable's effect, the regression models also provide estimates of the standard deviation, or standard error of the estimates. A rough rule of thumb, which is based on statistical assumptions about the form of the underlying error distribution in the model, is that when a regression coefficient is more than twice the size of its standard error (this ratio is called the *t*-statistic), the researcher can conclude that the effect of the explanatory variable is significantly different from zero.

When the outcome variable of interest is a discrete variable that can take on only the values of zero and one (such as staying or leaving), the multiple regression model must be modified to ensure that the estimates obtained are on average correct and to permit statistical testing of the hypotheses. There are various ways that this can be done. Both Thursby and Thursby and Black and Stephan have chosen to use a method called *logit analysis* because some of the outcome variables that interest them can only take on the values of zero and one. The estimated coefficient of an explanatory variable that comes from this model tells the researcher the impact of a one-unit change in the explanatory variable on the logarithm of the ratio of the probability that the dependent variable takes on the value of one to the probability that the dependent variable takes on the value of zero. These coefficients can be transformed, and the authors do so, to estimate the effect of a one-unit change in the explanatory variable, on the probability that the outcome will be a one.

Three of the chapters in this volume use panel data. Each has information for a number of years for a number of different observational units. Ehrenberg, Jakubson, and Rizzo have twenty-one years' data for each of 228 universities. Thursby and Thursby use a data set that has information on 3,342 science and engineering faculty members and that has 44,731 observations, for an average of thirteen years' data per faculty member. Finally, Borjas has data on foreign student and U.S. citizen student graduate program enrollments for six years at a number of American universities.

Panel data allows the researcher to control for variables that the researcher has no information about that may systematically vary across the units (universities in Ehrenberg et al. and Borjas and individual faculty members in Thursby and Thursby) but are constant for a given unit over time. If these unobserved variables are correlated with observed variables that are included in the model and the analysis does not control for them in the regression, the estimated associations between the explanatory variables and the outcomes that are found may, on average, be incorrect. One way to take account of these unobservable variables when the researcher has panel data is to estimate models that include a dichotomous (zero or one) variable for each unit in the estimating equation. When this is done, the estimation method is called a *fixed effects* model, and it is equivalent to estimating the original model in first difference form.

Notes

1. Much of the growth in industry's share occurred during the 1980s. Industry's contribution in percentage terms declined in the early years of this century, following the 2001 recession (National Science Board 2006, vol. 1, fig. 4.3).

2. The source of funding clearly affects the direction of university research. Just as defense funding has affected the university and the direction of its research (Leslie 1993), funding from business also affects the direction of university research as well as the culture of the university (Bowie 1994).

3. Owen-Smith (2003) argues that "increased patenting and commercial engagement on U.S. campuses" has "dramatically altered the rules that govern interuniversity competition" into a hybrid regime, where achievement in either the academic or commercial realm is dependent upon success in the other.

4. See, for example, Audretsch and Feldman (1996a, 1996b), Black (2004), and Jaffe (1989).

5. See, for example, Blumenthal et al. (1996), Bowie (1994), Eisenberg (1987), and Slaughter and Leslie (1997). Kleinman (2001, 227–28) notes that "even absent direct industry involvement in laboratory research, the norm of what Merton referred to as scientific 'communism' is, for a wide range of reasons, sometimes violated, and existing research suggests that it is a mistake for investigators to treat this ideal as strictly or typically adhered to."

FINANCING SCIENCE
AND ENGINEERING RESEARCH

1

Who Bears the Growing Cost of Science at Universities?

RONALD G. EHRENBERG, MICHAEL J. RIZZO,
AND GEORGE H. JAKUBSON

INTRODUCTION

Scientific research has come to dominate many American university campuses.[1] The growing importance of science has been accompanied by a growing flow of funds to universities to support research from federal and state government and corporate and foundation sources. What is not well known, however, is that an increasing share of the costs of the research at universities is being funded out of internal university funds. So it is natural for us to ask, who bears the growing cost of the internal funds spent on research at universities?

We begin in the next section by sketching the reasons for the growing cost of scientific research at universities and the reasons for the growing share of universities' research costs funded out of internal university resources.[2] The reasons for the latter include changes in federal indirect reimbursement cost policies and the growing cost of start-up funds for new faculty. We present evidence on the magnitude of start-up costs that universities face for new researchers in science and engineering fields from our survey (summer 2002) of department chairs, deans, and vice presidents of research at more than 200 public and private research universities.

Our chapter then turns to an estimation of who bears the costs of internal research expenditures. Using panel data for a twenty-one-year period for 228 research universities, we estimate the impact of growing internal university expenditures on research on student/faculty ratios and the substitution of lecturers for tenure-track faculty, on average faculty salaries, and on undergraduate tuition levels. Perhaps our most important findings are that universities whose research expenditures per faculty member from institutional funds have been growing the most rapidly in absolute terms, other factors held constant, exhibit the greatest increase in student/faculty ratios and, in the private sector, the largest increases in tuition levels. So while undergraduate students may benefit from being in close proximity to great researchers, they also bear part of the growing institutional costs of research in the form of larger class sizes, fewer full-time professorial rank faculty members, and higher tuition levels. While the magnitudes of these effects are quite small, they undoubtedly vary by field. To the extent that they are higher in science-based disciplines, they may help to explain the waning interest in science among domestic students, explored by others in this book. After all, closer proximity to a star does not necessarily mean greater access to a star, and students enrolled in labs and discussion sections staffed by graduate students and adjuncts may fail to find that the research experience provides the excitement necessary to propel them on to graduate school.

Finally, in the concluding section, we speculate on future directions that research on the impact of the growing cost of science on academic institutions might take and also about whether the growing efforts by universities to commercialize their faculty members' research may yield sufficient revenues to begin to offset the universities' increasing institutional costs of scientific research.

The Growing Importance and Costs of Science

Scientific research has come to dominate many American university campuses. Viewed in terms of 1998 dollars, the weighted (by faculty size) average volume of total research and development expenditures per faculty member across 228 American research and doctoral universities increased from about $70,000 per faculty member in 1970–71 to about $142,340 per faculty member in 1999–2000.[3] This growth in scientific research, which was fueled by the availability of funding from government, corporate, and foundation sources, derives primarily from the major progress being made in science and the importance of these advances to our society.

To take but one example, recent advances in decoding the human genome, in advanced materials and in information sciences, promise major progress in health-care treatment in the years ahead. Any university worth its salt wants to be a leader in these fields so that it can attract top faculty and undergraduate and graduate students, increase its research funding for its programs, and potentially achieve financial returns by commercializing its faculty members' research.

What is not well recognized, however, is that in spite of generous external support for research, the costs of research are being borne increasingly by the universities themselves out of institutional funds. During the 1970–71 to 1999–2000 period, the weighted average institutional expenditure on research per faculty member at the 228 universities more than tripled. As a result, the weighted average percentage of total research expenditures per faculty member being financed out of institutional funds rose from 11.2 percent to 20.7 percent during the period (Figure 1.1). Increasingly the academic institutions themselves are bearing a greater share of the ever-multiplying costs of scientific research.

It is natural for universities to think of their institutional expenditures for research as investments that will lead to greater academic accomplishments for their faculty, increased external research grant funding, and perhaps even enhanced revenue from the commercialization of their faculty members' research findings. Below, we report our efforts to test whether the last two beliefs have any validity. Nonetheless, the questions remain: why have institutional costs of research increased so much, and who bears their cost?

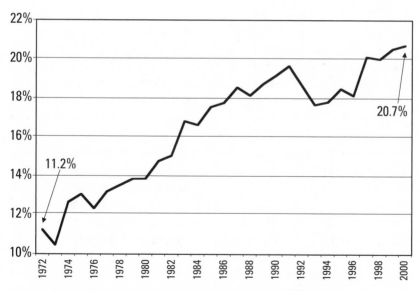

FIGURE 1.1 Weighted average percentage of total R&D expenditures by institutions, 1971–72 through 1999–2000.

Source: R&D data from NSF, Survey of Research and Development Expenditures at Universities and Colleges. Faculty data from IPEDS via WebCASPAR.

There are a number of forces that have led to the costs of research borne by universities to soar over the past three decades.[4] Theoretical scientists, who in a previous generation required only desks and pencils and paper, now often require supercomputers. Experimental scientists increasingly rely on sophisticated laboratory facilities that are increasingly expensive to build and operate. Research administration now includes strict monitoring of financial records and environmental safety as well as the detailed review and monitoring of experiments involving human subjects.

Historically, the federal government and other external funders, through the provision of indirect cost recoveries, have funded much of the research infrastructure that universities operate, as well as their research administration costs. Each institution was allowed to mark up the direct costs that its faculty members requested from external funders for research support by a multiple called the indirect cost rate, and the indirect cost revenues received on successful grant applications went to support the institution's research administration and infrastructure costs. However, after a well-publicized case involving Stanford University in the early 1990s, government auditors began to take a much harder look at universities' requests for indirect cost recoveries and put caps on the expenses that universities could claim in a number of areas (Kennedy 1997, 164–77). As a result, the average indirect cost rate at private research and doctoral universities, which was over 60 percent in 1983, fell to about 55 percent in 1997 and has remained near that level ever since.[5] On average, for any given level of direct cost research funding that their faculty members received, these private universities received 8.3 percent fewer funds from the federal government to support their research infrastructure and administrative costs in 1997 and thereafter than they did in 1983.

What is the likely response of an institution faced with such a reduction in external support for research infrastructure and administration? On the one hand, it might try to reduce its expenditures in these areas to match the decline in the external support for research. But such a strategy would alienate its faculty, who would view the institution's commitment to research as declining. In addition, if the reductions were made in areas in which the institution was not spending more than the maximum that the federal auditors would allow it to recover, the auditors would further respond by lowering its indirect cost rate in the following year. So, invariably private university administrators made up for the reduction in external funding to support their research administration and infrastructure costs by increasing their commitment of institutional funds.

In recent years the federal government has also placed increasing pressure on all universities to provide "matching" institutional funds for any research proposals that they submit. While universities try to provide matching funds out of resources that they would have spent for research even in the absence of a new external grant, they cannot always get away with doing this, especially for large-center grant proposals. Put another way, to compete for external funding, increasingly institutions have had to bear a growing share of the direct costs of their faculty members' research out of their own pockets.[6]

Finally, as scientists' equipment became more expensive and the competition for top-quality scientists intensified, the start-up funding that universities needed to provide to attract young scientists increased. Universities typically cannot recover these expenses in their indirect cost revenues billings, because the new scientists rarely have external funding when they arrive at the university. During the late 1990s, it was often alleged (although no systematic data supported the claim) that universities were providing young scientists in the range of $250,000 to $500,000 to set up their labs. The start-up costs of attracting distinguished senior scientists was said to be much greater. And even if these senior faculty members had federally funded research grants, these costs, too, were often not recoverable in indirect cost recovery pools because the institutions faced caps on their recoveries in a number of categories.

Because no systematic data on start-up costs had previously been collected, the Cornell Higher Education Research Institute (CHERI) conducted a "Survey of Start-Up Costs and Laboratory Space Allocation Rules" at research and doctoral universities during the summer of 2002. We surveyed three to six science and engineering departments at each of the research and doctoral institutions that we could identify (222 schools) using the 1994 Carnegie Foundation Classification of academic institutions. We surveyed the chairs of these departments, the deans of the colleges in which each of these departments were located, and the vice presidents or vice provosts for research in each university. In total, 1,031 department chairs, 408 deans, and 206 vice presidents or vice provosts received survey questionnaires.

These questionnaires began by describing things that are generally included as start-up costs (such as construction or renovation of labs, materials and equipment, support for laboratory staff, graduate assistants or postdoctoral fellows, summer salaries for the faculty member, reduced teaching loads, travel money, unrestricted research funding) and then asked respondents to provide information on the average and/or range of start-up costs incurred for new assistant professors and senior faculty members in their field, on the most expensive subfield in their discipline, and on the sources of funding for start-up costs.

Full results from this survey are available on the CHERI World Wide Web page (www.ilr.cornell.edu/cheri). Table 1.1 summarizes some of the information from the responses of the 572 department chairs (representing a response rate of 55 percent) that we received. The departments are grouped here into four broad areas—physics/astronomy, biology, chemistry, and engineering—and data are reported separately for private research, private non-research, public research, and public non-research universities.

Chairs were asked to report either the average start-up costs for faculty or the range of costs for faculty that they had experienced in the last year or two. Some reported both measures. Table 1.1 presents the average mean start-up costs that the chairs reported for new assistant professors and senior faculty as well as the averages for faculty in their most expensive

TABLE 1.1 Average mean start-up costs for departments by category.

		Private research		Private non-research		Public research		Public non-research	
		Average	Count	Average	Count	Average	Count	Average	Count
AA	PHYS	395,746	9	147,944	18	320,932	42	169,491	56
AA	BIO	403,071	14	199,754	26	308,210	38	172,582	55
AA	CHEM	489,000	20	221,052	29	441,155	43	210,279	71
AA	ENG	390,237	19	152,010	20	213,735	52	112,875	46
HA	PHYS	563,444	9	254,071	14	481,176	41	248,777	47
HA	BIO	437,917	12	208,886	22	430,270	37	217,082	49
HA	CHEM	580,000	17	259,348	23	584,250	40	284,269	60
HA	ENG	416,875	16	209,057	21	259,494	50	146,831	43
AP	PHYS	701,786	7	90,000	2	740,486	29	359,783	23
AP	BIO	957,143	7	481,458	12	651,087	23	438,227	31
AP	CHEM	983,929	14	532,046	11	989,688	32	550,349	33
AP	ENG	1,441,667	9	326,694	14	408,443	38	223,292	23
HP	PHYS	1,000,000	4	418,333	3	1,110,577	24	455,882	17
HP	BIO	1,575,000	5	555,500	10	856,250	16	709,444	27
HP	CHEM	1,172,222	9	575,000	8	1,187,115	26	648,913	23
HP	ENG	1,807,143	7	452,000	10	472,086	34	254,597	23

Source: Tabulation of responses to the Cornell Higher Education Research Institute's "Survey of Start-Up Costs and Laboratory Space Allocation Rules."

AA average start-up costs for new assistant professors   PHYS   Physics and Astronomy
HA high-end start-up costs for new assistant professors   BIO   Biology
AP average start-up costs for senior faculty   CHEM   Chemistry
HP high end start-up costs for senior faculty   ENG   Engineering

subfields; in each case the "count" represents the number of reporting departments in the category.[7]

This table suggests that, with few exceptions, at the new assistant professor level research universities provide larger start-up packages than other universities, and private universities provide larger start-up packages than public universities. Average start-up costs for assistant professors at private research universities in the four fields ranged from $390,237 to $489,000.[8] Estimates of the average high-end (most expensive subfield) assistant professor start-up cost packages at the private research universities varied across fields from $416,875 to $580,000.

Start-up cost packages for senior faculty members were considerably larger. For example, for the private research universities, the average start-up costs varied from a low of about $700,000 in physics to a high of about $1,442,000 in engineering. As at the assistant professor level, the packages at research universities tend to be larger than the packages at non-research universities, and the packages at privates tend to be larger than the packages at publics. However, at the senior level in a few cases start-up costs were larger at research publics than at research privates. This may reflect efforts by some publics to move to a higher level by hiring a number of key senior faculty members. In any event, start-up costs are clearly a major expense faced by American universities.

WHO PAYS FOR THE GROWING COSTS OF SCIENCE?

Does the Increasing Institutional Cost of Science Lead to Fewer Faculty, Lower Faculty Salaries, and Higher Tuition Levels?

How have universities responded to the growing importance and costs of science? One might suspect that the growing importance of science has provided an incentive for universities to allocate a greater share of their faculty positions or faculty salary dollars to scientists. However, using data from a set of arts and sciences colleges at leading private research universities, prior work by one of us concluded that over a recent twenty-year period, neither the share of faculty positions nor the share of the faculty salary budget devoted to scientists had increased at these colleges (Ehrenberg and Epifantseva 2001). Controlling for the growth of enrollments in the various disciplines or for whether overall faculty size was increasing or decreasing at each college did not alter these conclusions.

Of course it may well be that the increasing cost of science is felt throughout a university's budget. Colleges of Arts and Sciences may receive a declining share of their university's total faculty positions, even if enrollment changes do not warrant this loss, as more positions are allocated to science-intensive engineering and medical colleges. This particular hypothesis is difficult to test because some faculty in the latter two types of colleges, especially in medical colleges, are funded on soft money that they raise themselves.

More generally, to the extent that the other sources of income that a university receives, such as state appropriations, annual giving, and endowment income, are directed toward supporting an increasingly large scientific infrastructure, this may put upward pressure on undergraduate tuition or cause the university to cut back its expenditures on other areas. Inasmuch as the faculty salary bill represents a large chunk of institutional costs, it is possible that the increasing costs of science are distributed throughout the university in the form of slower rates of increase in faculty salaries and an increase in the student/faculty ratio above the level that

TABLE 1.2   Panel regression results (standard errors).

Explanatory variables	Student-faculty ratio*		Faculty salary†		In-state tuition	
	Private	Public	Private	Public	Private	Public
Internal funding per faculty, in $10,000	**0.535** (**0.139**)	**0.130** (**0.033**)	392 (107)	540 (47)	165 (45)	−11 (7)
Total giving per student, in $10,000	0.031 (0.039)	−0.005 (0.041)	134 (30)	−237 (58)	40 (13)	22 (8)
Endowment per student in $10,000	**0.007** (**0.003**)	**−0.039** (**0.009**)	31 (2)	18 (12)	4 (1)	2 (2)
State appropriations per student, in $1,000	−0.442 (0.460)	−0.041 (0.013)	119 (351)	92 (18)	−132 (149)	−9 (3)
Share of enrollments — graduates	**19.6** (**2.7**)	**−17.6** (**1.8**)	4484 (2097)	1455 (2522)	4822 (883)	1507 (368)
Within R²	0.161	0.105	0.887	0.823	0.897	0.785

Source: NSF, HEGIS, and IPEDS via WebCASPAR and Council on Aid to Education.

Notes:
All specifications include both fixed institution and year effects.
Significant at 95% level in BOLD.
All dollar values in $US 1998 dollars; data for AY 1977–98.
Seventy-eight private schools and 150 public schools in the sample.
* Calculated as FTE undergraduate and graduate enrollment per ranked faculty.
† Calculated as weighted average faculty salaries for all ranked faculty members (weighted by faculty size).

would otherwise prevail, all other factors held constant. It is to tests of these hypotheses that we now turn.

Table 1.2 uses data from a panel of 228 research and doctoral universities spanning the 1976–77 to 1997–98 period to explain why an institution's ratio of full-time equivalent undergraduate and graduate students to its full-time professorial ranked faculty, its average faculty salary level, and its undergraduate tuition and fee level (in-state for publics) vary over time.[9] The multivariate regression technique used in the analysis is discussed in the statistical methods section of the introduction of this volume.

The explanatory variables are the institution's research expenditures per professorial ranked faculty out of its own internal funds, the share of its enrollments that are in Ph.D. and nonprofessional master's programs, the level of contributions per student that it received during the year from all sources, its endowment per student, and its state appropriation per student. All variables that are expressed in dollars are measured in constant (1998) dollars.

The contributions, endowment, and state appropriations variables are included in the model to control for other sources of revenue coming into the university that can be used to support faculty positions and faculty salaries. Another major source of institutional revenue, its undergraduate tuition and fee level, is omitted as an explanatory variable because it is treated as an endogenous outcome, and suitable instruments do not exist that would permit us to identify a structural model that treated tuition as endogenous.[10] The share of the institution's enrollments in Ph.D. and nonprofessional master's programs is also included as a control variable to capture the impact of changes in the degree mix of students on the outcomes.

All equations include institutional fixed effects, and, as a result, our estimates indicate the impacts of changes over time in the explanatory variables on student/faculty ratios. They also include fixed year effects to capture the effects of omitted macro-level variables that may influence the outcomes.[11] Separate equations are estimated for public and private universities.

Our key findings are found in the first row of Table 1.2. Turning first to the student/faculty ratio equations, other factors held constant, universities whose institutional expenditures on research per faculty member are increasing the fastest in absolute terms are also the ones whose student/faculty ratios are increasing the fastest. The magnitude of the relationship is greater for private universities than it is for publics, and we cannot reject the hypothesis for the former that each $10,000 increase in institutional research expenditures per faculty member is associated with an increase in the student/faculty ratio of close to 0.5. During the period, the weighted average real institutional research expenditure per faculty member at the private universities in the sample increased from about $7,700 to $17,500. So, on average, the increase in institutional research expenditures per faculty member at the privates has caused an increase in student/faculty ratios of about 0.5 during the period, as compared to what the student/faculty ratios would have been if the increase in institutional research expenditures per faculty member had not taken place.

The magnitude of the relationship is somewhat lower for the public university sample, 0.130. However, the growth in absolute terms of real institutional research expenditures per faculty member has been larger for the public universities; the weighted average for public universities in the sample rose from about $7,600 to $31,300 during the period. Hence, the impact of the increase in public universities' expenditures on research out of their own funds on student/faculty ratios has probably been somewhat smaller, increasing, other factors held constant, by about 0.3 during the period.

Our estimates of the impact of changing internal expenditures on research on the student/faculty ratio prove to be robust to a number of specification changes. Including full-time equivalent student enrollments on the right-hand side of the equation did not alter the finding. When we used five-year averages for each institution to capture longer-run changes, we found larger student/faculty ratio effects, although they tended to be less statistically significant because of the reduction in our sample sizes.

When we repeated the analyses, using total full-time faculty (including lecturers and instructors) in the denominator of the student/faculty ratio rather than professorial ranked faculty, similar positive coefficients on the institutional research expenditure per faculty member variable were obtained. However, the magnitudes of these coefficients were somewhat smaller than the coefficients reported in the first row of Table 1.2. Hence, while the data suggest that an increase in institutional research expenditures per ranked faculty member probably does lead to some substitution of full-time lecturers and instructors for professorial ranked faculty, the ratio of students to all faculty (including the lecturers and instructors) also increases when institutional expenditures on research per ranked faculty member increases.

Turning next to the average faculty salary equations, we find evidence that increasing institutional research expenditures per faculty member leads, other factors being equal, to higher (not lower) faculty salaries. This may reflect reverse causality—high-paid faculty members with strong research records being recruited and requiring considerable institutional research funding for start-up cost (see the previous section).

Turning finally to the in-state tuition equations, our estimates suggest that private universities that increase their institutional research expenditures per faculty member, other factors held constant, also experience larger increases in their undergraduate tuition levels. However, the magnitude of this relationship is quite small. Given that the weighted average expenditures on institutional research expenditures per faculty member at private universities in the sample increased by about $10,000 in real terms over the period, our estimates suggest that undergraduate tuition levels at these institutions averaged about $165 higher in real terms in 1998 than they otherwise would have been if institutional research expenditures per faculty member had remained unchanged during the sample period. This increase represents less than 1 percent of the average tuition and fee level of more than $20,000 that prevailed at selective private research universities in 1998, and this modest increase in tuition needs to be balanced against the gains that undergraduate students at research universities experience from being educated in close proximity to leading researchers.

While our major interest is in the impacts of increasing institutional research expenditures per faculty on the various outcomes, several other variables' effects are worth mentioning. First, we find that as the share of Ph.D. and nonprofessional (research) master's students increases at these universities, undergraduate tuition levels also increase, other factors held constant. During the period, these shares rose on average from 0.240 to 0.306 at the private research universities and from 0.145 to 0.178 at the public research universities. Our estimates thus imply that tuition levels at the private research universities were about $336 higher and those at the public research universities about $50 higher at the end of the period, in real terms, than would otherwise have been because of the growth in the share of students that were enrolled in Ph.D. or nonprofessional (research) master's programs.

Hence, other factors held constant, undergraduate students do bear some of the cost of increased Ph.D. and research master's program size in the form of higher tuitions.[12] However, again the magnitude of this effect is quite small. In addition, since increasing the share of students enrolled in these programs appears to attract better faculty to universities, as measured by the higher average faculty salaries that result when these program shares expand Ph.D. program shares (significantly higher in the case of the private universities), undergraduate students may also benefit from increases in the relative size of these programs.

Statistically we also find that increased annual giving per student and increased endowment levels per student are both significantly associated with higher undergraduate tuition levels. Higher levels of endowment and annual giving provide increased subsidies for students, increase the attractiveness of the university to students, and hence should increase the tuition levels that students are willing to pay to attend the university.[13] Increases in state appropriations per student are negatively associated with tuition increases at public universities, which is consistent with the findings of previous research (Rizzo and Ehrenberg 2004).

Sensitivity Analyses

We conducted two different types of sensitivity analyses.[14] First, we tested whether the relationships differed between research and other universities in the sample. Second, we sought to ascertain whether the estimated effects of the growth of institutional research expenditures per faculty members differed when we specified a more complex model in which changes in annual giving, in research expenditures per faculty member out of external funds, and in state appropriations were all treated as endogenous.

When we estimated separate equations for research and other institutions, we found that a given increase in institutional research expenditures per faculty member has a larger effect on student/faculty ratios at the research institutions. Changes in an institution's commitment of institutional funds to research have the largest positive association with average faculty salaries at the public non-research institutions, the institutions that probably need to offer higher salaries to attract more research-oriented faculty. Finally, it is at the private non-research universities that increases in institutional research expenditures per faculty member have the largest marginal effect on tuition levels.

To develop a more complex model, we first specified an equation in which an institution's annual change in external research expenditures per faculty member depended on the institution's annual change in institutional research expenditures per faculty member (through the route of start-up costs, matching funds, and more general research infrastructure support), the changes in the national research budgets that different government agencies are receiving, and the shares of the institution's external research funding it received from the various agencies during the previous year. So, for example, holding the change in institutional research funding constant, institutions that derive a greater share of their external research budget from NIH should see greater increases in their external research funding than institutions that derive a greater share of their research funding from NSF, when the NIH research budget rises relative to the NSF research budget.

This external research funding equation was important to us because previous research has shown that at private research universities, higher levels of total research spending per faculty member are associated with higher levels of annual giving (Ehrenberg and Smith 2003). Hence, treating annual giving as an exogenous source of revenue that does not depend on the institution's research volume may bias our results. So we next specified an equation for change in annual giving that depended on the change in total research expenditures per faculty member at the institution. Finally, some states explicitly or implicitly reward institutions for increasing their volumes of external research funding, so treating state appropriations as exogenous may also bias our results. Hence, we also specified an equation for change in state appropriations per student that depended on an institution's changes in external research volume per faculty member.

In the context of this more complex model, a change in internal research expenditures per faculty member will directly affect each of our three outcomes and also indirectly affect them through its effect on the change in external research expenditures, which in turn may affect state appropriations per student and (along with its change in institutional research expenditures per faculty member) may also effect annual giving per student.

When we estimated this more complex model and then simulated what the effects of changes in institutional research expenditures per faculty member were on the student/faculty ratio, average faculty salaries, and annual tuition levels, our results were very similar to the ones presented above. Treating external research funding, annual giving, and state appropriations as endogenous did not change any of our findings.

CONCLUSION

Our research suggests that undergraduate students bear at least part of the increased costs that American universities are undertaking for their faculty members' scientific research in the forms of higher student/faculty ratios, some substitution of lecturers for professorial rank faculty, and, in private universities, higher tuition. However, the magnitudes of these effects are surprisingly (to us) small. Whether these costs are more than offset by the benefits the students receive from being educated in proximity to scientific researchers who are at the cutting edge of their disciplines is an open question that deserves serious study.

Our research also suggests that as the share of Ph.D. students and non-professional master's students increases at both the public and private research universities, undergraduate tuition also increases at these universities, other factors held constant. While Gordon Winston's (1999) work indicates that no undergraduate student at a major university bears the full cost of his or her education, our results suggest that the undergraduates bear a part of the cost of graduate education, in the sense that the average subsidy (in Winston's terms) they receive from attending the university is lower because of the presence of more graduate students. Again, whether these costs are more than offset by the benefits the students receive from being educated close to (and sometimes by) graduate students is an open question deserving of serious study.

Our study has only begun to touch on the impact that universities' growing institutional expenditures on research has had on the universities. The relatively small impact that we observed on student/faculty ratios may mask decreases in scientists' teaching loads, which in turn may have led to larger class sizes and greater use of part-time faculty members. The growth of science may have also crowded out other things. For example, increased institutional support for teaching or research assistants for scientists and for stipends for graduate students in the sciences may have led to decreased availability of funds to support teaching assistants or graduate students on fellowships for humanists and social scientists or to slower rates of growth of graduate student stipends in these fields.[15] Or, to take another example, increased institutional support for scientific research facilities and start-up costs for scientists may have reduced the funding that otherwise would have been available for travel and other "perks" in the humanities and social sciences.

It is possible, of course, that the increasing costs of research that are borne by universities may be eventually at least partially offset by revenues that the universities receive from increased commercialization of their faculty members' research. The Association of University Technology Managers (AUTM) reported in their fiscal year 2000 survey of their members that American colleges and universities received more than $1 billion in licensing income and other forms of royalties relating to patents. While this figure seems large, it was concentrated in a few large "winners"; 90 percent of the universities in their sample received less than $2 million that year, and almost half received less than $1 million.[16]

Licensing income received in one year depends on the flow of investments in research that universities have made in the past. If we ignore this and the fact that the return on any particular research project may occur for a number of years in the future, a simple way of looking at the commercial returns that universities receive from their faculty members' research is to ask how the licensing income received by a university in one year relates to its own expenditures on research during that year. Licensing income received in fiscal year 2000 averaged 3.23 percent of *total* research expenditures for the year across the institutions in the AUTM sample. As we have noted, universities fund about 20 percent of research expenditures out of their own resources, which suggests that licensing income averaged about 16 percent of institutions' research expenditures out of *internal* university funds during the year.

At first glance this seems like a significant return, but this calculation is misleading for at least three reasons. First, the licensing income that universities receive is divided between the university and the researchers. So, only a share of the revenue actually comes to the university itself. Second, focusing on the average ratio ignores the skewness in the distribution of research returns. The median institution in the sample licensing income was 0.83 percent of its total research revenue, which is about 4.2 percent of its internal volume of research expenditures. Third, given the volume of a university's research, licensing income and other forms of revenue from patents that are related to this research do not fall off trees. Rather, they must be "harvested." Considerable efforts must be made by universities and their faculty members to decide if faculty members' discoveries have potential commercial value, to patent the discoveries, to seek partners to develop commercial potential, to negotiate licenses or equity positions, and to enforce patents.[17] All of these activities require resources. Indeed, the cost of trying to enforce patents alone can prove very expensive.[18]

While no comprehensive source of data on the costs that universities incur in trying to generate licensing income is currently available, summary information from the AUTM licensing survey permits us to make some back of the envelope calculations. During fiscal year 2000, the 142 U.S. universities in the AUTM sample employed a total of 479.95 "licensing" full-time equivalent employees (FTEs) and 494.53 other FTEs in their technology transfer offices. They also incurred $117,927,842 in legal fees, of which third parties reimbursed them only $53,685,716.[19] Hence, these universities' net legal fees for technology transfer activities were roughly $64 million, and they employed a total of about 975 employees, including patent attorneys, other professionals, and support staff. If we assume that the fully loaded costs of each employee (salaries, benefits, office space, etc.) averaged $100,000 that year, the total expenses of technology transfer activities for these institutions were in the range of $161.5 million dollars, or an average of about $1.15 million per university.

Maintaining the assumption that the average fully loaded cost of each employee was $100,000, the AUTM survey responses allow us to compute an estimate of the net licensing income (income after expenses) for 138 of the universities in the sample. The mean net licensing income in this sample was $6,554,200, but the median was only $343,952. By our calculations, 51 of the 138 institutions actually lost income that year on their commercialization activities, and we estimate that the median net licensing income for the 87 that made money was $1,309,828. When one remembers that the licensing income received by universities is split between them and the faculty members whose patents have generated the income, it seems clear that commercialization of research has yet to provide most universities with large amounts of *net* income to support the universities' scientific research activities.

Notes

1. At least partially in response to this phenomenon, the Association of American Universities launched an initiative to reinvigorate the humanities in America's major research universities (Mathae and Birzer 2004).

2. See Brainard (2005) for a similar discussion.

3. The figures that follow are all computed from the NSF WebCASPAR system (http://caspar.nsf.gov).

4. These forces are discussed in more detail in Ehrenberg (2000, chap. 6) and Ehrenberg (2003).

5. Indirect cost rates at the public research and doctoral universities were lower at the start of the period and actually rose slightly. The lower initial rates were due to many publics not having to return funds that the state had spent constructing new research facilities. The increase came about because declining state support for operating budgets of public higher education made it more important for the public universities to try to tap all available potential sources of revenue (Ehrenberg 2000, 2003).

6. However, in October 2004, after many years of discussion, the National Science Foundation eliminated the cost-sharing, or matching funds, requirement for specific projects for which they solicited research (Field 2004).

7. When only a range was reported, we used the midpoint of the range to calculate for the average.

8. Due to the large variability of the size of start-up cost packages across institutions in each field, the differences that we observe between fields (here and below) are often not statistically significant.

9. Professorial ranked faculty members include assistant, associate, and full professors.

10. Results that we obtained for the impact of changes in institutional research expenditures per ranked faculty member on faculty/student ratios and average faculty salaries are very similar to results obtained when tuition is included in the model and treated as endogenous.

11. For example, increases in the unemployment rate may increase students' financial need, which in turn will put pressure on institution's financial aid budgets, influencing the salaries that it can afford to pay its faculty, its student/faculty ratio, and its tuition level.

12. We should stress, in keeping with the views of Gordon Winston (1999), that we are not saying that undergraduate students are subsidizing Ph.D. students. Rather, all students attending the research universities receive subsidies, and growing the relative size of Ph.D. programs simply reduces the size of the monetary subsidy that undergraduate students receive.

13. Again, see Gordon Winston (1999). Empirical evidence that in-state and out-of-state tuition levels are positively associated with levels of endowment per student at public universities is also found in Rizzo and Ehrenberg (2004).

14. Details of these analyses are found in Ehrenberg, Rizzo, and Jakubson (2003).

15. It is not surprising that the leaders of the growing movement to unionize graduate assistants on private university campuses tend to be graduate students from the humanities and soft social sciences (Ehrenberg et. al 2004).

16. See Blumenstyk (2002a). Some of these large winners were universities that cashed in equity positions that they had taken in companies in lieu of receiving licensing income. Blumenstyk (2003b) presents similar data for fiscal year 2002. Slightly less than $1 billion was received in that year, and two-thirds of the total went to just thirteen institutions. The institutions in the AUTM survey vary from year to year, so the data are not strictly comparable across years.

17. Thursby and Thursby (2000) describe this process in much more detail and provide estimates of licensing production functions.

18. The University of Rochester has established an "eight figure" legal fund in its effort to obtain billions of dollars in royalties from the makers and marketers of the arthritis drug Celebrex (Blumenstyck 2002b). Recently, a federal district court judge ruled against the university, and the case is now under appeal (Blumenstyck 2003).

19. Association of University Technology Managers (2001), attachment D.

2

How Does the Government (Want to) Fund Science?

Politics, Lobbying, and Academic Earmarks

JOHN M. DE FIGUEIREDO AND BRIAN S. SILVERMAN

How Is Academic Scientific Research Funded?

The United States' "system of innovation" is often touted as one of the most productive in the world. Many consider academic research—much of which is funded through the federal government—to be a key pillar supporting that system (Mowery and Rosenberg 1993). Over the past twenty years, the federal government has funded roughly 25 percent of all academic research. In fiscal year 2003 alone, the federal government appropriated more than $22 billion to the scientific endeavors of universities (Shackelford 2004), distributing the funds chiefly through the National Institutes of Health (65 percent), the National Science Foundation (12 percent), the Departments of Defense (8 percent) and Energy (3 percent), and NASA (4 percent).

Federal funds for university research are, generally speaking, allocated through two mechanisms. The first, and most common, is a competitive allocation mechanism, which accounts for nearly 90 percent of the federal funds. The most well known of the competitive processes is "peer-review" selection in which scientists submit proposals for specific research projects to federal agencies. These proposals are reviewed by experts in the relevant fields, whose evaluations enable the agencies to rank proposals based on perceived scientific merit. The widespread practice reflects Vannevar Bush's strong conviction more than sixty-five years ago that scientists would be most productive if left to decide for themselves how "government support should be distributed and how their efforts should be organized" (Dickson 1998, 25).

Although agencies may incorporate other criteria into their decision calculus (such as whether a particular project will likely yield a result particularly useful to the defense of the country), the peer-review process is generally seen as a mechanism for allocating scarce funds for the most promising scientific research efforts (Nelson and Rosenberg 1993). Despite this, critics have raised questions about the benefits of the peer-review system. Chubin and Hackett (1990) summarize much of this, saying that "expert evaluation is inevitably plagued by cronyism, elitism, and conflicts induced by self-interested competition" (165). Moreover, prominent legislators and academics have argued that the peer-review process effectively serves to concentrate research funding among a few elite schools whose scientists populate the peer-review boards (Gray 1994). In the eyes of critics, the peer-review process also tends to reward "safe" research projects that conform to accepted beliefs, thus starving truly breakthrough research (Silber 1987).

In recent years, and partly as a response to these types of critiques, a second means of allocating funds has assumed increasing importance. Known as "academic earmarking," this second method is a political process that entirely bypasses the peer review described above. Academic earmarking is the process by which legislators place specific provisions in the government's annual appropriations bills requiring specific agencies to allocate specified levels of funding to designated universities for particular projects (Savage 1999). As part of the federal appropriations bills, these earmarks become law.

In fiscal year 2003, the U.S. budget included 1,964 academic earmarks, accounting for more than $2 billion, or just over 10 percent of all federal funding for academic research (Brainard and Borrego 2003).[1] More striking is the rate at which academic earmarking has grown. Since 1980, earmarks have increased 5,900 percent in real-dollar terms, representing a cumulative annual growth rate (CAGR) of 19.4 percent, accelerating in recent years to a CAGR of 31 percent since 1996. In contrast, since 1980 overall federal funding of science has experienced a 220 percent growth, or CAGR of less than 4 percent. Together, these data demonstrate that academic earmarking accounts for an increasing share of the total federal research budget for universities and that the rate of its share capture has increased substantially over the past decade.

The rise of earmarking has sparked a significant debate among universities, policy makers, and political economists as to the wisdom of this approach. Many in academe have decried the growth in earmarks, fearing that increased earmarking will inevitably cut into the amount of federal funding for research that will be allocated through the peer-review process.

These same critics often lament the desire of elected officials to steer money toward politically beneficial projects, which may not coincide with projects of high scientific potential. Supporters have argued, however, that politicians, not peer-review boards, can best represent the needs and long-term interests of the country and taxpayers. As noted above, these supporters contend that the peer-review system does not allocate resources optimally and that only radical change can help middle- and lower-tier schools compete with those who have, for years, been ensconced in the top tier.

This study sheds light on this debate by examining the evidence related to several key questions: Are academic earmarks distributed differently than competitive funds? If so, then what determines the allocation of earmarked funds, and how productive are earmarked research grants? In answering these questions, we first examine the supply side of earmarks. We show that a university's political representation is a significant predictor of whether that university receives academic earmarks. Indeed, members of the House and Senate Appropriations committees send a disproportionate amount of academic earmarks to their home districts.

We then explore the demand side. It is here that results are somewhat striking. While some earmarks do simply "appear on the doorstep" of universities, a large portion of academic earmarks are directed to a university at the university's request. That is, universities proactively lobby their political representatives in Washington for academic earmarks, and these representatives then deliver on the request. Moreover, universities that profess to eschew earmarks actually, at times, lobby and receive these exact same earmarks they claim to avoid.

The study then examines the literature on the quality of research that emanates from peer-reviewed and earmarked research projects. The few studies to date suggest that earmarked funding leads to research with lower impact than does competitive funding and that universities that receive earmarked funding at one point in time do not subsequently improve their research standing. Finally, we explore what this all means for the future of science funding. Namely, we discuss the pro-active role of universities in seeking academic earmarks and what advantages and disadvantages such a system may provide.

The Distribution of Academic Earmarks

Figure 2.1 presents data on the success of medium-sized Carnegie I universities at obtaining federal research funding through each of the two channels described: peer-review grants and earmarked grants. If allocation decisions in these channels are driven by similar processes, or based on similar university capabilities, then we would expect a strong positive correlation between the amounts of money received through each funding channel. A cursory glance at Figure 2.1 shows no obvious positive correlation between funding levels through competitive grant processes and funding levels through academic earmarks. Indeed, a correlation analysis confirms that these two funding levels are not correlated (σ = -0.01).

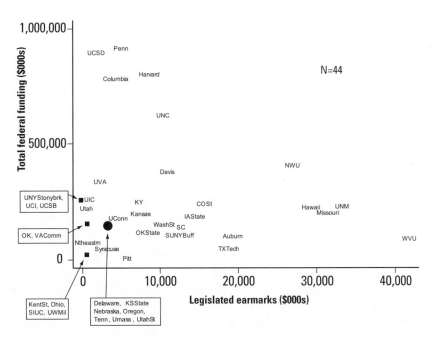

FIGURE 2.1 Federal money for medium-sized Carnegie I universities.

Source: NSF and Center for Responsive Politics.

Notes:

Carnegie I research universities with enrollments of 17,000 to 26,500 students for which data is available.

All data is cumulative for 1997 to 1999.

Total federal funding is the total government research dollars awarded to the school from all government agencies (e.g., DOD, DOE, NSF, NASA, USDA, etc.). Almost all money is granted through a type of competitive grant-making process. This data is obtained from the National Science Foundation.

A legislated earmark is the amount of funding legislated by Congress in appropriation bills. This data is obtained from the Center for Responsive Politics.

A closer look at the figure reveals that several of the universities that fared particularly well between 1997 and 1999 at obtaining legislative earmarks were located in the states and districts of senators and representatives who had strong leadership roles in the U.S. Congress, notably West Virginia University (Robert Byrd, ranking member, Senate Appropriations Committee); University of New Mexico (Jeff Bingaman, Senate deputy Democratic whip; ranking member, Energy and Natural Resources Committee); University of Hawaii (Daniel Inouye, second-ranking Democrat, Senate Appropriations Committee); and University of Missouri (Richard Gephardt, House minority leader). This suggests that the political power of a university's legislators may play a role in the allocation of academic earmarks.

Supply of Earmarks

The U.S. Congress has a long tradition of legislators directing money to their districts. Historically, this has manifested itself in farming subsidies, highway grants, and other infrastructure projects. According to both the public choice theory in political economy and popular belief, such actions help legislators enhance their reelection chances by enabling them to claim credit for creating local jobs and otherwise bringing to their districts a "fair share" of government largesse. Universities, however, are relative newcomers to this game.[2]

Despite the relatively recent entry of universities into this arena, there is both qualitative and quantitative evidence that earmarks are often supplied because legislators see it as in their interest to secure such services for constituent universities. Qualitatively, Savage (1999) recounts an instance in which John Murtha (D-PA), a member of the House Appropriations Committee, wrote into an appropriations bill an earmark from the Department of Defense for Marywood College, a small Catholic college in his district, that the school did not request and for which it had no obvious use (Savage 1999, 133). Quantitatively, Table 2.1 offers statistical evidence

TABLE 2.1 Means in receipt of earmarks, FY 1997–FY 1999.

For all universities in districts with and without Appropriations Committee members

	Average earmark
No Senate Appropriations Committee member (n = 3442)	$144,693
Senate Appropriations Committee member (n = 3704)	$313,686
No House Appropriations Committee member (n = 6131)	$187,331
House Appropriations Committee member (n = 1015)	$503,839

concerning the unconditional means of academic earmarks allocated during the 1997–99 period. This table shows that the average earmark awarded to a university whose senator served on the Senate Appropriations Committee (SAC) was more than double that of a university whose senator did not serve on this committee. The average earmark awarded to a university whose representative served on the House Appropriations Committee (HAC) was higher still, roughly triple that of a university without such committee representation.

Figure 2.2 revisits the legislative earmarks obtained by medium-sized Carnegie I universities. In this figure, however, university identities are replaced

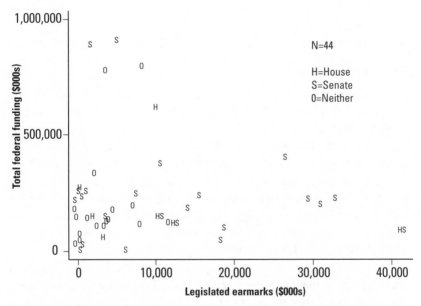

FIGURE 2.2 Federal money for medium-sized Carnegie I universities based on political representation.

Source: NSF and Center for Responsive Politics.

Notes:
Carnegie I research universities with enrollments of 17,000 to 26,500 students for which data is available.

All data is cumulative for 1997 to 1999.

Total federal funding is the total government research dollars awarded to the school from all government agencies (e.g., DOD, DOE, NSF, NASA, USDA, etc.). Almost all money is granted through a type of competitive grant-making process. This data is obtained from the National Science Foundation.

A legislated earmark is the amount of funding legislated by Congress in appropriation bills. This data is obtained from the Center for Responsive Politics.

by information on the HAC and SAC representation enjoyed by each school. Two features stand out from this figure. First, the vast majority of schools that receive substantial earmarks have HAC or SAC representation. Second, Senate representation appears to be more heavily associated with earmarks than does House representation; 52 percent of schools in this sample with SAC representation received more than $10 million in earmarks, whereas 43 percent of schools in the sample with HAC representation received more than $10 million in earmarks. As a point of comparison, 100 percent of schools with both HAC and SAC representation received more than $10 million in earmarks, while only 6 percent of schools with no HAC or SAC representation received earmarks greater than $10 million. The evidence in the figure (and in Table 2.1) suggests that academic earmarks have become much like other earmarks—a mechanism for legislators to funnel discretionary federal spending to their districts in precise and directed ways.

Demand for Earmarks

Institutions of higher learning have hardly been passive recipients of earmarks. Rather, universities have dramatically increased their efforts to lobby Congress specifically to obtain earmarked funds. In recent times, this effort has undoubtedly been hastened by the increased costs that universities have assumed for research (discussed by Ehrenberg, Rizzo, and Jakubson in the previous chapter). Although university lobbying dates back to well before World War II, the modern practice of lobbying for earmarks began in the late 1970s. In 1978, Tufts president Jean Mayer, keen to build a nutrition center at the university, sought out lobbyists Gerald Cassidy and Kenneth Schlossberg to press his case with their contacts in Congress. Cassidy and Schlossberg ultimately succeeded in getting into a 1979 appropriations bill a line stipulating that the Department of Agriculture should give Tufts University $32 million toward costs of building a new nutrition center (Savage 1999, 102). Perhaps more importantly, this effort led Cassidy and Schlossberg and other universities to recognize the opportunity to lobby for the federal funding of directed academic earmarks. By the late 1990s, such lobbying had developed into a cottage industry. Nearly 300 universities lobbied during at least one year between 1997 and 1999; most of these hired professional lobbyists from one of a handful of firms dedicated to lobbying for academic earmarks, while others also maintained a staff of internal lobbyists for this purpose. These universities spent $160 million on lobbying during this time period (de Figueiredo and Silverman 2006).[3]

To be sure, not all university lobbying is directed at obtaining earmarks.

A small number of "elite" universities lobby for science policy (for example, increased budgets for the National Science Foundation or for rules that will facilitate stem cell research). Indeed, many elite schools deny that solicitation of earmarks is an integral part of their lobbying campaigns. However, a recent sunshine law, the Lobbying Disclosure Act of 1995 (and its attendant technical amendments in 1996), requires all organizations that spend more than $20,000 in a given year on lobbying to disclose information about these lobbying efforts. One can review these lobbying reports for evidence of the locus of lobbying efforts. Three things are clear from such a review. First, middle- and lower-tier schools direct virtually all of their lobbying efforts and expenditures toward academic earmarks. Second, even those elite schools that decry the practice of academic earmarking often devote at least some lobbying effort to seeking earmarks. Third, the degree to which elite schools are engaging in lobbying and obtaining earmarks has been steadily increasing over time.

When the phenomenon of academic earmarking first appeared in the 1980s, the Association of American Universities (AAU)—a trade association of sixty-two elite research institutions that account for the lion's share of competitively awarded federal research funding—publicly called for the cessation of such earmarks. But over time, the AAU was less able to present a united front on this issue. By the late 1980s, a number of AAU member schools began to pursue earmarked funding, and AAU members engaged in often-rancorous debate about the propriety of accepting and soliciting earmarked funds. Despite several attempts by individual schools to lead efforts to collectively refuse to pursue or accept earmarks, in fiscal year 2003, 90 percent of the AAU membership (fifty-four of the sixty U.S. members) accepted at least one earmark.[4] In 2003 AAU members received a total of $336 million in earmarks, representing 21 percent of all earmarked funds.

Table 2.2 revisits the statistical evidence concerning academic earmarks that was first presented in Table 2.1, but it includes information on lobbying expenditures by universities. The table shows the statistics for all universities and also for the "lobbier" (lobbying expenditures > 0) subsample. In the full sample, the results show that the average university with no representation on the SAC spent $9,430 lobbying and received an earmark of $144,693, for an unconditional average return of roughly $15 for every $1 spent on lobbying. However, universities with representation on the SAC lobbied about 40 percent less than their nonrepresented counterparts yet received a little more than two times the earmark, for an unconditional return on investment of almost $56 for every $1 spent on lobbying.

Universities with representation on the HAC obtained an unconditional return on investment of almost $25 for each $1 spent on lobbying. Results are similar when the analysis is limited to those institutions that actually lobby.

While the static unconditional means provide the first glimpse of an interesting story, we can gain additional insight by examining Appropriations Committee "switchers." Although there are few legislators who rise to or leave the Appropriations Committees in any year, there were four switchers on the Senate Appropriations Committee after the 1998 election. In that election, Senator Lauch Faircloth (R-NC) was defeated by Senator John Edwards (D-NC), while Senator Dale Bumpers (D-AR) retired. Both senators thus lost their positions on the SAC, and neither North Carolina nor Arkansas had another senator on the committee. Senators Jon Kyl (R-AZ) and Richard Durbin (D-IL) were elevated to the SAC to replace the two outgoing senators, becoming the only senators from Arizona and Illinois to sit on the committee. Table 2.3 provides the data for lobbying expenditures and earmarks for the four states affected by these switches. As the table shows, there was a large jump in lobbying by North Carolina universities in the year after Faircloth's exit, but the earmarks to North Carolina universities shrank by half in that year. Arkansas universities similarly increased their lobbying expenditures after Bumpers's exit but saw an increase

TABLE 2.2 Means in lobbying and earmarks, FY 1997–FY 1999.

For all universities in districts with and without Appropriations Committee members

	Average lobbying expenditures	Average earmark
No Senate Appropriations Committee member (n = 3442)	$9,430	$144,693
Senate Appropriations Committee member (n = 3704)	$5,595	$313,686
No House Appropriations Committee member (n = 6131)	$7,414	$187,331
House Appropriations Committee member (n = 1015)	$7,612	$503,839

For lobbying universities in districts with and without Appropriations Committee members

	Average lobbying expenditures	Average earmark
No Senate Appropriations Committee member (n = 247)	$131,410	$1,157,920
Senate Appropriations Committee member (n = 176)	$117,750	$2,987,555
No House Appropriations Committee member (n = 363)	$125,225	$1,477,928
House Appropriations Committee member (n = 60)	$128,765	$4,588,803

in earmarks. Table 2.3 also shows that after Kyl's ascension to the SAC, Arizona universities did not change their lobbying level but did experience a 41 percent increase in earmarks. Durbin's ascension was followed by an increase in both lobbying and earmarks. Thus, in three of the four cases of committee switchers, there is evidence that both lobbying and earmarks respond to changes in SAC membership.

The above data might suggest that there is an enormous payoff to lobbying. However, Tables 2.2 and 2.3 report unconditional means. It is unclear from these tables how large the return to lobbying is after controlling for other factors. While political scientists and economists have explored lobbying extensively, both theoretically (Austen-Smith 1993, 1995; Rotemberg 2002; Ainsworth 1993) and empirically (Tripathi et al. 2002; de Figueiredo and Tiller 2001; Wright 1996) there has been, until recently, no successful attempt to measure the returns to lobbying in a large-scale statistical study. This has been largely due to data limitations. But de Figueiredo and Silverman (2006), taking advantage of information revealed under the Lobbying Disclosure Act and of features of academic earmarks that facilitate surmounting other empirical challenges, demonstrate that universities that lobby Congress receive dramatically higher earmarks than their nonlobbying counterparts—*if* the lobbying university is in the state or district of an Appropriations Committee member.

De Figueiredo and Silverman (2006) estimate the elasticities of university lobbying to academic earmarks. Figure 2.3 depicts their results graphically. Notably, a 1 percent increase in lobbying expenditures by a university without representation on the House Appropriations Committee (HAC) or Senate Appropriations Committee (SAC) results in a 0.15 percent increase in earmarks, but in many specifications the point estimates are not significantly different from zero. In contrast, if the university is represented by a member of the HAC or SAC, then this 1 percent increase in lobbying yields

TABLE 2.3 Senate Appropriations Committee switchers: Unconditional means for universities in their states.

	Faircloth R-NC (n = 109)		Bumpers D-AR (n = 33)		Kyl R-AZ (n = 23)		Durbin D-IL (n = 109)	
	Lobbying	Earmarks	Lobbying	Earmarks	Lobbying	Earmarks	Lobbying	Earmarks
1997 ($)	1,481	148,863	0	104,863	12,608	294,129	10,688	139,961
1998 ($)	3,425	164,895	0	63,037	14,782	295,275	9,555	162,745
1999 ($)	7,222	86,112	606	141,261	14,347	405,523	11,666	174,93

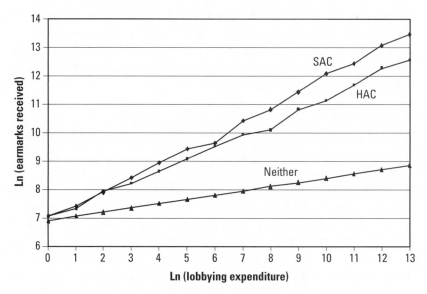

FIGURE 2.3 Effect of Lobbying Expenditure on Earmarks received.
Note: The "Neither" category is not statistically significant from zero.

a 0.28 percent or 0.35 percent increase in earmarks, respectively. These latter elasticities are statistically different from zero.[5] As indicated by the different Y-intercepts in Figure 2.3, de Figueiredo and Silverman also find that, after controlling for lobbying, the main effects of HAC and SAC representation increase the average earmark by 16 percent and 19 percent, respectively. Thus, consistent with the previous subsection's emphasis on the supply of earmarks, HAC and SAC representation both result in money being sent to the university in the absence of lobbying.

In sum, universities get earmarks because they ask them, although some earmarking would persist even without lobbying due to legislators' supply-side pressures. The above-cited statistical results suggest, and we conjecture, that the active lobbying by universities for academic earmarks is partially responsible for the high and increasing level of earmarking.

EARMARKS TO RESEARCH

Having discussed the supply and demand for earmark funding by universities, a natural question arises: Does academic earmarking lead to different research, or research of differing quality, than peer-reviewed fund allocation?

A commonly raised concern about academic earmarking is that by distributing research funds to politically connected institutions rather than those with the most competitive research proposals, academic earmarks will lead to the funding of lower-quality research than will peer review (Brainard and Borrego 2003). De Figueiredo and Silverman (2006) find that, after controlling for SAC/HAC representation and lobbying, schools that receive earmarks tend to be lower-tier research universities (as ranked by the National Research Council).[6] In addition, the results also indicate that earmarking redistributes funds away from top schools toward lobbying schools with powerful political representation.

The fact that, on average, earmarks are bestowed on lobbying universities with Appropriations Committee representation and on lower-tier schools is not prima facie evidence that earmark-funded research is of lower quality than that supported by peer review. As noted above, some have argued that the competitive grant process systematically underrates promising projects from lower-tier schools. Although it is unlikely that this bias would systematically affect states and congressional districts of Appropriations Committee members, the practice of earmarking could partially rectify such a bias by funding promising projects proposed by lower-tier institutions and by constituents of Appropriations Committee members. In such circumstances, the research produced by earmark-supported research should be no less productive than competitive grant-supported research. Alternatively, it is possible that earmarked funding to lower-tier schools today will provide resources that enable these schools to become higher-tier schools and thus better able to compete in the peer-review competition tomorrow. One can attempt to examine these claims empirically.

There are two tests that can help provide insight regarding the benefits of this redistribution. First, do earmark-funded projects result in significantly better or worse research than peer-reviewed projects? Unless earmark-funded research is measurably less useful than its peer-reviewed counterpart, concerns about earmarking could be unfounded. Second, do earmarks provide an opportunity for lower-tier schools—which otherwise might not be able to obtain a large amount of peer-reviewed funding—to "catch up" with top-tier universities? If earmarking enables schools to improve significantly over the long term, then it may be justifiable even in the face of weaker research in the short term.

A number of recent studies have attempted to compare the importance of research funded by peer review and by earmarking and set-asides. Payne (2002, 2006) finds that set-asides and earmark-funded research generates more papers per research dollar than peer-review–funded research. However, the citation rates of these papers are statistically and substantially lower than that of peer-review–funded projects.[7] The author interprets this as evidence that earmark-funded papers have significantly less impact on subsequent scientific advance. Though the quantity of output is higher, the quality of output is lower when earmarks are present.

Evaluating the long-term changes in university quality that can be attributed to earmarking is difficult given the only recent widespread diffusion of academic earmarking. However, earlier studies of the relationship between a university's receipt of earmark funds and its subsequent change in academic ranking have generally found no systematic effect of earmark funds on ranking movement over the past fifteen to twenty years (Savage 1999). One may argue that the indicators of importance may be noisy and that the results—which are based on aggregate university research productivity—would be different if analyzed on a project-by-project basis. Nevertheless, the initial finding across these studies is not favorable to earmarking. It is clear that more research is needed in understanding the productivity of earmarking.

Conclusion:
How Do We (Want to) Fund University Research?

The earmarking process is now pervasive. It is clear that some of the critiques of the peer-review process are addressed through earmarking. In particular, academic earmarking does result in a redistribution of federal research money away from top-ranked schools to middle- and lower-tier schools. Earmarks are granted to projects that are unlikely to be funded through peer review, and, moreover, the money is often allocated to construction projects, which, along with overhead, serve to maintain the infrastructure of the university.

However, these benefits are not without costs. The evidence suggests that the redistribution of money is highly influenced by political representation. That is, there is a strong redistribution effect away from universities without senior representation on the relevant appropriations committees and toward those universities with just such representation. This move to the political arena means that universities move resources into private-interest lobbying for these earmark grants. Whether lobbying for earmarks is a public good or a public bad is an open question. The current evidence, however, suggests that not only does earmarking crowd out peer-review competitive grants, but it also results in lower-quality research (as measured by citation rates).

This raises the important public policy question: How do we want to fund science? It seems that the funding of science has a tipping point. To the extent that there is a nearly one-for-one substitution of earmarks for peer-reviewed funding,[8] there is cause for concern that, at some point, the amount of funding allocated via earmarks will reach a "tipping point" after which we will see a rapid, wholesale shift from peer-reviewed funding to earmarked funding of most academic research. Thus far many elite research schools have not pursued earmarks with full intensity and continue to devote the bulk of their lobbying efforts to increasing the size of the federal budget for academic research. But as earmarks consume more of this budget and peer-reviewed funding concomitantly less, at some point it will no longer be in the interest of the elite research schools to go after the peer-reviewed funds because the pot of money will be simply too small.[9] Indeed, we already see that 90 percent of AAU universities receive 21 percent earmark funding. If earmarking becomes a sufficiently large proportion of the academic research funding pie, then all universities will begin lobbying for earmarks, and the federal financing of academic research may end up resembling highway appropriations, where nearly every dollar is earmarked to a particular roadway project.

Fundamentally, there is a pernicious collective action problem at work here (Savage 1999). Although it is better for the university research system as a whole to fund research through peer-review systems, it is individually rational for a single university below some threshold level of research quality to seek out an academic earmark. However, since this earmark reduces the level funding available for competitive grants, the next bottom school that pursues competitive funding finds it increasingly difficult and shifts its resources from competitive review to political lobbying for earmarked funds. Left unchecked, this process continues to raise the threshold below which universities seek earmarks until the peer-review system eventually unravels. While the system can likely withstand some amount of earmarking without completely unraveling, the substantial and quick rise of earmarking seems to be pushing us closer to a tipping point.

Ultimately, the question that must be answered collectively is "How do we want to fund science?" As the rise of earmarking shifts the basis of competition for research money, there are many questions that we should ask. Do we want the direction of science to be determined by elected officials? Will the new basis for competition enable lower-tier institutions to improve their research capability, and, if so, is this the best way to achieve that goal? And ultimately, can we maintain the extraordinary system of innovation in the United States if we move toward earmarking more and more federal science dollars?

We seem to be on the slope of increasing acceptance of, and resignation to, the earmarking of federal funds for research. We can alert policy makers to the advantages and disadvantages of this type of allocation of federal research dollars, but it is ultimately up to the politicians (suppliers) and university presidents and their lobbyists (demanders) to set the course for the future.

NOTES

1. *The Chronicle of Higher Education* collects and cleans data on these earmarks. We use the *Chronicle*'s data in the statistical studies described below. Data available at http://chronicle.com/stats/pork/ (registration required).

2. The debate over academic earmarking thus fits within the broader literature on federal discretionary spending and congressional committee structure. In this literature there are conflicting results as to whether representation on a committee results in that committee spending more of the committee budget in committee members' districts than in non–committee members' districts. A small group of studies finds that committee members direct spending into their districts (e.g., Ferejohn 1974; Hall and Grofman 1990; Plott 1969), while a large number of studies find no effect (e.g., Mayer 1991; Ray 1980). Others have found mixed results (e.g., Anagnoson 1980; Arnold 1981).

3. Brainard (2002) finds that the total is approximately $90 million. We have been unable to reconcile these figures with our own.

4. The AAU member institutions in the United States that did not receive an earmark in FY 2003 are Caltech, Duke, University of Chicago, SUNY Stony Brook, Washington University in St. Louis, and Yale. Of these six schools, all but Caltech received at least one earmark in either FY 2001 or FY 2002.

5. De Figueiredo and Silverman (2006) discuss the details of the average and marginal returns to $1 of lobbying.

6. In addition, universities with medical schools are more likely to receive earmarks, ceteris paribus. This may reflect the entrepreneurial nature of medical school faculty at pursuing earmarks (Richard Levin, personal communication, July 22, 2002, NBER Summer Institute meeting), or it may reflect the political attractiveness of medical research.

7. Although citations are a common measure of quality, there are a number of reasons to question the metric as a measure of quality. See Kleinman (2005, 87–88).

8. One university president related to us his discussion with a high-ranking

official in the Bush administration who told him that if universities would stop receiving earmarks, then the administration would put each and every dollar back into the peer-reviewed pot of money for universities.

9. The debate over earmarking thus also fits into the broader literature on "good government" and the effect of rent-seeking on productivity. Murphy et al. (1993) demonstrate that rent-seeking behavior is subject to increasing returns, suggesting that an initially small amount of rent-seeking behavior can spiral upward toward a high rent-seeking equilibrium. High levels of rent-seeking efforts can "crowd out" other, more productive efforts. One mechanism for stemming such behavior is for a government to commit to "high quality" policies that effectively preclude giving in to rent-seeking parties (La Porta et al. 1999). Viewed through this lens, the evidence to date indicates that scholars have significant cause for concern regarding academic earmarking and its potential effect on academic research in the U.S.

The Commercialization
of Science

3

University Science Research Funding: Privatizing Policy and Practice

RISA L. LIEBERWITZ

INTRODUCTION

Contemporary discussions of the university often seek to balance its traditional public role with the increasing expansion of university-industry relationships. Advocates of university commercial activities and "alliances" with corporations argue that they are necessary to finance costly research programs that will benefit the public as well as foster economic growth. Yet critics of university privatization and commercialization argue that increasingly close university-industry relationships compromise the values of academic freedom and independence that are fundamental to the university's institutional identity and social role and change relationships among university faculty. This study addresses these issues in the context of life sciences research, which has emerged most recently as the university program with large needs for financial support and with great commercial potential. The chapter begins with a background of the changes in law and public policy that have encouraged and facilitated closer relations between universities and industry in the life sciences, leading to a dramatic growth in university patenting and licensing practices and corporate financing of university science programs. The next section of the chapter examines the critiques of these changes in policies and practices. The final section explores the range of proposals for legal and policy reforms to address potential or existing conflicts created by the growth of university market activities.

CHANGES IN SCIENCE, PUBLIC POLICY, AND UNIVERSITY PRACTICES

The culture of academic science traditionally has reflected deeply held values that promote openness and the sharing of research through scholarly publications. These values of "communalism" have been fundamental to the advancement of science, enabling scientists to build on the research foundations laid by their colleagues (Barber 1953; Merton 1973).[1] A closely related value is "disinterestedness," which equates scientists' self-interest with the public interest to ensure that research contributes to the public good of expanding knowledge rather than to a narrower personal interest (Eisenberg 1987, 183–84; Rai 1999, n.67). Moreover, scientists' independence is required to enable them to freely choose their research agendas and engage in uninhibited debate and critique (Rai 1999, 91). These communal norms have not suppressed competition among scientists, who are known for their fierce competitiveness to make discoveries ahead of their colleagues (92). The structure of the university science reward system reinforces scientists' social role of contributing to the public interest while also providing individual recognition (Eisenberg 1987, 183–84). As the first to publish significant research results, an individual scientist achieves public recognition of his/her excellent research and gains public respect and status (183). This recognition and respect is enhanced through full disclosure of research methods and data, which permits other scientists to confirm the results through replication of the research.

These values are consistent with the traditional norms of academic freedom that provide faculty independence from influence or pressure by third parties, including governments, boards of trustees, or corporate financial donors (Rai 1999, 92–93; Lieberwitz 2002, 80–85). Such independence has been identified as essential to fulfilling the goal of the university to serve the public good through teaching and research based on the search for truth and the expansion of knowledge free from conflicts of interest.[2] Academic freedom of individual faculty includes the right to choose research agenda, pursue research on controversial matters, determine course content in teaching, and express public opinions on extramural matters (Lieberwitz 2005, 111–12). Professional academic freedom is also a collective right, including the autonomy of the profession to engage in self-governance through the process of peer review (Lieberwitz 2002, 80–85).

Although university scientists have deeply internalized these communal values, key developments in science, politics, and law have converged to promote values and practices in conflict with the traditional culture of academic science. These developments have had an especially strong impact on the life sciences, where academic research in genetics and biotechnology exploded in the mid-1970s with the discovery of recombinant DNA (rDNA) technology, described as "the single pivotal event in the transformation of the 'basic' science of molecular biology into an industry" (Kenney 1986, 13–15, 23, 32–33).[3] In this new age of biotechnology, the line between basic and applied research has been increasingly blurred, as the potential commercial value of basic research becomes clearer at an earlier stage and as the time period between basic research and its application is shortened (Kenney 1986, 106; Mowery et al. 2004, 185).[4] Donald Stokes (1997), seeking a new typology for research that is motivated both by a quest for fundamental knowledge as well as for use, describes such research as occurring in "Pasteur's Quadrant."

Political and legal developments of the 1980s expanded the potential for university market activities, consistent with U.S. public policy emphasizing privatization of governmental functions through measures such as contracting out of public services (Kenney 1986, 28–29). Although federal funds from agencies such as the National Institutes of Health (NIH) continued to comprise the greatest percentage of funding for university science research programs,[5] congressional appropriations for research funding began to drop in the 1980s (Kenney 1986, 28–29; Eisenberg 1987, n.2). At the same time, Congress promoted university commercial activities by enacting the 1980 Bayh-Dole Act,[6] which authorized and encouraged federal fund recipients, including universities, to apply for patents on results of federally funded research.[7] Prior to the Bayh-Dole Act, federal law had granted the government title to inventions developed with federal funds, a policy that favored placing these inventions in the public domain. The government agency could choose to dedicate the invention to the public domain by publishing the results without obtaining a patent or by providing nonexclusive licenses to private parties seeking to use a government-owned patent (Eisenberg 1996, 1675–76; Rai 1999, n.113). In the 1960s and 1970s, universities could patent and license federally funded academic research in cases where the funding agencies—most notably the Department of Defense and to a lesser extent the National Science Foundation and the Department of Health, Education, and Welfare—transferred title in the research to the universities (Mowery et al. 2004, 45, 51–52; Eisenberg 1996, 1683–84, 1691–92).

Congress justified the Bayh-Dole Act as a means of commercializing publicly funded research results, with particular emphasis on promoting collaboration between universities and industry (35 U.S.C. sec. 200).[8] Congress reasoned that if a commercial business had exclusive rights to a federally funded invention, as either patent-owner or as licensee of a university-owned patent, the business would be more willing to invest the large amounts of research and development funds needed to develop a marketable product

(Eisenberg 1996, 1698–99). Permitting the government contractor to patent the federally funded research would also avoid the bureaucratic complexities of navigating among twenty-six different federal agency regulations in the process of applying for licensing rights (Eisenberg 1996, 1663–64, 1676; Eisenberg 1987, 181–82; Mikhail 2000, 378).

Developments in the courts further strengthened the commercial potential of academic research. Almost simultaneously with the enactment of the Bayh-Dole Act, the U.S. Supreme Court held, in *Diamond v. Chakrabarty* (1980, 447 U.S. 303), that life forms can be patented, in this case a genetically engineered bacterium that degraded crude oil. The Court agreed that "products of nature, whether living or not" are not patentable but concluded that "human-made inventions," including living materials, may be patented (303, 313). With this decision, the Court expanded the impact of the Bayh-Dole Act by opening the door to patent applications from universities and businesses conducting life science research, including "smaller pieces of DNA that are significantly [far] away from practical commercial application" (Rai 1999, 104). Biomedical patents constitute close to 50 percent of university patent activity (Rai and Eisenberg 2003, 54). Private biotechnology companies have sought patents on basic research tools, including DNA sequences, gene fragments, and cell receptors (Heller and Eisenberg 1998, 699).[9]

Universities responded positively to the legislative and judicial developments expanding the potential for patenting federally funded research results, as shown by the increase in university patents after the effective date of the Bayh-Dole Act.[10] In 1979, before the Bayh-Dole Act, U.S. universities obtained 264 patents, compared with 1997, when American universities obtained ten times that number, 2,436 patents (Rai and Eisenberg 2003, 53). In fiscal year 2000, U.S. universities filed for 8,534 patents, an increase of 12 percent over the prior year (Blumenstyk 2002c). From 1980 to 1990, patent applications on NIH-funded inventions increased by almost 300 percent (Krimsky 1999, 22). From 1991 to 2000, the patents granted to U.S. universities increased by 131 percent (Blumenthal 2003, 2454), and licenses granted by the universities increased by 158 percent (2455). Between 1988 and 2003, U.S. patents awarded to academic institutions quadrupled, from about 800 to more than 3,200 per year.[11] This change in patent activity is also a reminder of the importance of federal funding for university research programs. Since the post-WWII period, federal funding has consistently been the most important source of university research financial support, ranging from approximately 60 percent to 70 percent of university research funding since 1960 (Kenney 1986, 35–36; Press and Washburn 2000, 39; Slaughter and Rhoades 1996, 327, Mowery et al. 2004, 24). Under the Reagan administration in the 1980s, federal science research funding was cut significantly, though the cuts were ultimately less drastic than feared (Kenney 1986, 28; Krimsky 1999, 20). Public funding has maintained its importance for university life sciences research, estimated at 70 percent to 80 percent of total funding for university life sciences research (Blumenthal 2002, 380).

At the same time that Bayh-Dole was encouraging commercialization of publicly funded research, corporate financing of university research programs also grew significantly, increasing by 93 percent between 1980 and 1984 (Eisenberg 1987, n.2). In 1994, 90 percent of life sciences companies had a relationship with an academic institution, including a significant increase in industry funding of life science academic research (Blumenthal et al. 1996, 371–72), estimated at $1.5 billion (11.7 percent) of the $12.8 billion of all extramural support of life science academic research (371).[12] Though most (71 percent) life sciences research projects are funded by industry at less than $100,000 a year and are relatively short in duration (371), "a considerable number" are large, long-term projects funded at $500,000 or more per year, including funding for basic research (369). University-industry agreements provide for the terms of the exchange, including the corporation's right to exclusive or nonexclusive licensing rights to any university-owned patents resulting from the research program supported by corporate funds. As the corporate funding becomes more significant, the likelihood increases that the corporation will negotiate for exclusive licensing rights and a significant presence in the university (Dueker 1997, 498).

At the largest scale of industry funding, often referred to as "strategic corporate alliances," a single corporation provides tens of millions of dollars to finance an entire academic department or research program in exchange for exclusive licenses, influence over research agendas, and access to the university facilities, faculty, and graduate students (Kenney 1986, 55–72). For example, under the controversial and much-publicized 1998 agreement between the Swiss pharmaceutical corporation Novartis/Syngenta and the University of California, Berkeley, Novartis/Syngenta agreed to provide $25 million over five years to UC–Berkeley's Department of Plant and Microbial Biology in exchange for exclusive corporate licensing rights to about a third of the department's discoveries. Additionally, UC–Berkeley gave Novartis/Syngenta two of five seats on the department's research funding committee, the right to review all proposed publications and presentations by participating faculty and their graduate students, and the right to ask for a ninety-day publication delay to provide time for patent applications (Press and Washburn 2000, 41–42; Elliott 2001, 21; Blumenstyk 2001).[13]

Despite the expansion of their technology transfer offices, universities have not yet reaped substantial profits from patenting and licensing, with most universities using such revenue to cover the costs of running the technology transfer activities (Eisenberg 1996, 1713–14; Mowery et al. 2004, 190). Further, a small number of universities account for most of the revenue from technology transfer activities (1713–14).[14] Even successful university technology transfer operations are identified as comprising only 1 or 2 percent of their universities' total research budget, with the most successful ones relying "on single 'blockbuster' patents for the majority of their revenue" (Dueker 1997, 466; Johnston 2007, 162–63). One commentator, though, views the royalty income to universities to be "a significant and valued source of revenue" in spite of the fact that it is small compared to the amount of federal funding of university research.[15]

Increased commercialization of academic research has made an impact on both individual faculty and university research programs. As different academic disciplines have expanded from basic research to applied research programs with commercially marketable research products, faculty have been attracted by the potential for profits. In the context of publicly funded research, the Bayh-Dole Act (sec. 202[c][7][c]) requires the university to share with a faculty member the profits from royalties related to a university-owned patent invention created by that faculty member. Today, in the life sciences, faculty in fields such as plant and human genetics are engaged in research of interest to industry, including agribusiness and pharmaceutical companies.

Faculty often enter into consulting arrangements with businesses while continuing their university teaching and research, with an estimate that about half of life sciences faculty act as consultants for industry (Blumenthal 2002, 379; Krimsky 1999, 19). Since the mid-1980s, 21 to 28 percent of life sciences faculty have consistently received research support from industry (378). During that time period, about 7 to 8 percent of faculty reported that they hold equity in a company related to their research (379). During the 1980s and 1990s, faculty participated in founding twenty-four Fortune 500 companies and more than 600 non–Fortune 500 companies in the life sciences (385).

CRITIQUES OF COMMERCIALIZATION OF ACADEMIC RESEARCH

The Public Costs of University Patents

The Bayh-Dole Act initiated a major change in university practices by permitting federally funded recipients to patent the results of publicly funded research. This legislative change has been hailed by supporters as promoting the public interest by making academic research more readily available to industry through licensing of university patents (Thursby and Thursby 2003, 1052; Ramirez 2004, 365–66).[16] Thus, this focus on the market shifts attention from the public domain as the repository of academic research to the private marketplace as an equally valid or even superior destination for faculty research results.[17] Moreover, this legislative encouragement of patenting publicly funded research indirectly encourages increased commercialization of privately funded academic research. If it is legitimate to patent publicly funded research and license those patents to industry, then patenting of privately funded research and exclusive licensing to corporate sponsors may appear uncontroversial. Closer scrutiny of this premise, though, reveals serious costs to the university and the public of commercializing academic research, whether it is publicly or privately funded.

Patents impose social costs by removing academic research discoveries from the public domain (Eisenberg 1996, 1667–68). The monopoly control given to the patent holder during the patent period is generally justified as promoting the public good by providing an incentive to individuals to risk the investment of capital to invent and then to disclose their inventions to the public (1668). The patent system's incentives to invent and disclose are unnecessary for university researchers, though, whose professional norms already create such incentives (Eisenberg 1987, 183–84; Rai 1999–2000, 88–94). Consistent with the university's public mission, academics traditionally make such disclosure by placing their research developments directly into the public domain through scholarly publications (Eisenberg 1996, 1663; Rai 1999–2000, 92–93; Krimsky 1999, 39). By permitting universities to patent their federally funded inventions, the Bayh-Dole Act creates a double cost to the public, as the public initially pays for federally funded university research and pays again when the federally funded invention is removed from the public domain under a university-owned patent (Eisenberg 1996, 1666).

The Bayh-Dole Act's privatizing of publicly funded research imposes additional costs on the public by undermining the public mission of the university. Universities are supported by public funds and tax exemptions,

with the expectation that they will serve the public interest through education and research activities independent of university self-interest or private
third-party interests (Blumberg 1996, 101–5). In contrast, by encouraging
universities to engage in patenting and licensing of federally funded research
results, Bayh-Dole creates a conflict between these private commercial
interests and university's institutional mission to engage in independent
research in the public interest (Lieberwitz 2004, 782). The public pays financially for these conflicts of interest. Through exclusive licensing agreements,
a business can reap the benefit of publicly funded academic research by,
in effect, purchasing the university's patent.[18] Although the industry licensee
may also invest more capital and take further risks in developing products
that use the university-owned patent, the public funds that covered the
cost and risks in developing the original patent still provide a significant
indirect public subsidy to the business. The public pays multiple times—
for the federal grant to the university, for the removal of the patented
invention from the public domain, and for the higher prices that a business may charge where it has a monopoly use over the university-owned
patent.[19] Perhaps even more costly to the public is the less tangible cost
resulting from the loss of the university's independence from outside interests, undermining its institutional legitimacy.[20]

Patenting of privately funded research results in some of the same public costs as the Bayh-Dole Act. Corporations that provide funds for academic research also reap the benefits of the publicly supported university
infrastructure that makes it possible for faculty to carry out privately
funded research. Further, patenting and licensing of privately funded academic research still has the same effect of constricting the public domain.
The same institutional conflicts of interest also exist, as the university's
public mission and independence are compromised by university-industry
agreements exchanging corporate funds for exclusive licenses to university-
owned patents. Charles Caldart has described the relationship: "In a very
real sense, the universities are now experiencing a shift from corporate *contribution* to corporate *investment* in academia" (1983, 25). Public perception of university research will also be negatively affected, as the public no
longer can trust in the independence of academic research from corporate
influence.

Given the increase in corporate funding of academic research, the conflicts of interest are a serious issue at both individual and institutional levels of the academy. Universities typically agree to delays of publication for
periods ranging from three to six months, or even longer, to provide time

for a corporate funder to review research results and for a patent application to be filed prior to disclosure through scholarly publication (Eisenberg 1987, 216–26; Krimsky 1999, 30; Newberg and Dunn 2002, 192–93).[21] Although universities defend such practices as a reasonable process to protect any corporate trade secrets disclosed by the corporation to the researcher, such pre-publication review places the corporation in a privileged and potentially influential position in relation to the research. There have been incidents of corporate pressure to change research reports to eliminate negative results in relation to a corporate product.[22] Studies have reported that corporately financed researchers are significantly more likely to reach favorable results concerning a corporation's product, including pharmaceutical products (Press and Washburn 2000, 42; Cho and Bero 1996, 485; Clayton 2001, 11; Krimsky 2003, 142–49). The scientific journals *Nature, New England Journal of Medicine, Journal of the American Medical Association,* and *Lancet* now require authors of articles accepted for publication to "submit sources of funding, records of employment, and histories of financial investments."[23] Yet even these disclosures do not reveal the practice of some pharmaceutical companies to commission university professors to write an article, which is then ghostwritten by an employee of the drug company or medical marketing company and submitted to the professors for their approval prior to submission to a medical journal.[24]

"Strategic corporate alliances" have taken the conflicts of interest to the institutional level, with private corporate financing on the scale of tens of millions of dollars to support entire departmental research programs. The agreement between the university and the corporate funder will normally include the corporation's right to exclusive licensing of any university patents resulting from the funded research program and the corporation's active presence at the university (Press and Washburn 2000, 41–42; Elliott 2001, 21; Blumenstyk 2001). Although such strategic corporate alliances are not as widespread as individual faculty research contracts with industry,[25] their presence demonstrates the extent to which universities have become closely aligned with corporate commercial interests (Washburn 2005, 1–24). Such university-industry partnerships undermine the university's public mission and independence by giving commercial businesses a powerful position in shaping the research agendas of entire departments. The independence and legitimacy of the research findings is also called into question, given the conflicts of interest created by the level of dependence on a particular corporate sponsor (Slaughter and Leslie 1997; Etzkowitz et al. 1998). An external evaluation of the large-scale corporate funding agreement between UC–Berkeley and Novartis/Syngenta included extensive discussion of concerns about conflicts of interest and concluded that universities should avoid such comprehensive agreements covering entire departments (Busch et al. 2004, 142–47, 152).

The Effects of Patents on the Conduct of Science

The commercialization of university research has compromised the traditional values of communalism, disinterestedness, and independence of science (Blumenthal et al. 1996, 372–73; Eisenberg 1988, 1375; Kenney 1986, 108–11, 121–31). University scientists report changes in their relationships with colleagues engaged in research financed by private corporations: they are unwilling to discuss their research methods or results either because of corporate funding contracts that restrict the researcher from sharing the information or that give the corporation the right to see this information first or because of the researcher's interest in protecting information for future patents (Blumenthal et al. 1986, 1361–66; Krimsky 1999, 29–31).[26]

As Daniel Kleinman has argued, intellectual property law and practice have broadly and pervasively shaped academic biology research, including regular consideration by university scientists of commercialization of research through patenting, changes in academic culture, and impediments on access to scientific knowledge (2003, 114–37). The combined effect of the commercial potential of basic research in the life sciences and the relaxed patenting standards applied by the courts increases the impact of university patenting, which can cover a broad scope of basic research tools that can then be licensed to industry (Cripps 2004, 4–12; Heller and Eisenberg 1998, 699–700; Rai and Eisenberg 2003, 54–56). As Kleinman has noted, control over research tools affects power relations in science, making academic researchers dependent on corporations that own patents or licenses on research tools, as well as on other university researchers' willingness to share basic research with each other through "material transfer agreements" (2003, 124–26, 136–37).

The proliferation of patents has restricted the free use of basic research tools, including DNA sequences and gene fragments, needed to make further research advances in the biomedical sciences (Heller and Eisenberg 1998, 699–700; Rai and Eisenberg 2003, 55–56). Michael Heller and Rebecca Eisenberg (1998, 698) describe this "privatization of upstream biomedical research" as "the tragedy of the anticommons," which may discourage scientists from engaging in downstream research requiring multiple licenses on patented research tools. Further, the owner of a patented upstream research tool may seek to share in the profits of subsequent downstream patents through licensing agreements that provide for royalties on future sales of downstream products or for exclusive or nonexclusive licenses to use downstream discoveries.[27] Such "reach through" patent rights may thus inhibit future investment in research that depends on using patenting tools. Even if the future research is carried out, reach through rights will increase litigation over distribution of profits from the later invention (Heller and Eisenberg 1998, 699–700; Rai and Eisenberg 2003, 55–56; Kleinman 2003, 124–25). A well-publicized example of such reach through agreements is found in the use of the genetically engineered "oncomouse," which had been developed by Harvard University scientists with funding from DuPont Corporation.[28] Harvard patented the oncomouse and exclusively licensed the patent to DuPont, which, in turn, conditioned sale of the oncomouse on reach through rights to share in the profits from any products developed with the use of the oncomouse (Heller and Eisenberg 1998, 699–700; Murray 2006, 24–25).[29]

The U.S. Supreme Court recently sidestepped a potentially monumental decision on the scope of patents on basic scientific knowledge. In *Laboratory Corporation v. Metabolite Laboratories* (*LabCorp*),[30] the federal circuit court had upheld a patent that included natural phenomena[31] and the scientific test developed through use of such basic scientific knowledge. Specifically, the patent covers both the biological correlation between high levels of the amino acid homocysteine and deficiencies in vitamin B and folic acid, as well as the medical test to measure the homocysteine levels.[32] These deficiencies can result in serious or life-threatening conditions, including vascular disease, birth abnormalities, and cancer (Griffin 2006, C1), but, once diagnosed, can be treated easily with vitamin B and folic acid supplements (LCML 2006, 1). The Supreme Court had granted a writ of certiorari to review the federal circuit court's decision. Having received the briefs of the parties, the U.S. solicitor general, and twenty *amicus curiae*, and having heard oral arguments,[33] the Court dismissed the case, concluding that the writ of certiorari had been "improvidently granted." Justice Breyer, joined by Justices Stevens and Souter in a lengthy opinion dissenting from the dismissal, castigated the Supreme Court majority for its unjustifiable refusal to decide whether the patent "improperly seeks to 'claim a monopoly over a basic scientific relationship.'"[34] The dissenters concluded that the correlation between high homocysteine levels and vitamin B and folic acid deficiencies is an "unpatentable 'natural phenomenon.'"[35] Resonating with Heller and Eisenberg's anticommons theory, Justice Breyer

explained that patenting of basic scientific principles "can discourage research by impeding the free exchange of information" through overly broad
restrictions of the public domain and the resulting costs of obtaining
licenses for use of patented natural phenomena.[36] As Justice Breyer commented during oral arguments in the case, patenting every "useful idea"
could establish "monopolies in this country beyond belief."[37] With the
Supreme Court's dismissal of the case, the federal circuit court's decision
stands, providing precedent for approving patenting of natural phenomena.

   LabCorp is significant as well in revealing the impact of the Bayh-Dole
Act in promoting universities' active role in expanding the commercialization of academic research. At the core of the case is publicly funded
academic research in the 1980s, which was patented by the University of
Colorado and Columbia University (LCML 2006, 1–2). The universities
assigned the patent to a for-profit patent management firm, Competitive
Technologies, which then sublicensed the patent to Metabolite Laboratories,
a small for-profit business established by the University of Colorado (LCML
2006, 6–7).[38] LabCorp, a large for-profit clinical testing company, paid royalties for use of the test until it began using a newer test from another
manufacturer (7–8). Competitive Technologies and Metabolite Industries
sued LabCorp for patent infringement. The universities thus resolved their
conflicts of interest in favor of the market, regardless of the negative effects
that could result for the free flow of scientific research.

   Universities' commitment to protecting their patents in the world of commerce may undermine their ability to claim an "experimental use" exemption for the use of patents in the context of academic research. The common
law experimental use exemption is a judicially created exemption from
patent law enforcement for research carried out for "idle curiosity" or as a
"strictly philosophical inquiry."[39] The experimental use exemption, though,
provides less protection for university research than might be assumed by
many academic scientists in light of the courts' narrow definition of the
exemption (Burchfiel 1995, 351–53). The federal circuit court recently narrowed the exemption even more, holding that universities and nonprofit
organizations in general do not enjoy any greater access to the experimental use exemption than do for-profit businesses. In *Madey v. Duke
University* (307 F.3d, 1351 [Fed. Cir. 2002]) the court found that universities, in using patented inventions to carry out their "business" of scholarly
research, are not acting within the narrow experimental use exemption
(Madey 2002, 1362; Rai and Eisenberg 2003, 55). As the court additionally
noted, despite Duke University's claim to have used patented research tools

in furtherance of a noncommercial research program, Duke also engaged in "an aggressive patent licensing program from which it derives a not insubstantial revenue stream" (Madey 2002, n.7). After denying certiorari in *Madey*, the Supreme Court, in *Merck KGaA v. Integra Lifesciences I., Ltd.* (2005, 545 U.S. 193), expanded the patent law statutory research exemption[40] for preclinical drug development research, creating the ironic result of providing broader research exemptions to the pharmaceutical industry than to basic scientific academic research.[41]

RECOMMENDATIONS

Proposed reforms of public policy and private practices seek to address the problem of the restricted public domain of biomedical research discoveries. These proposals fall into categories of reforms of law, federal agency policies and practices, and voluntary patenting and licensing practices by universities and industry. Some of these reforms have been partially adopted, while some would require new measures.

Proposed legal reforms call for changes in judicial doctrine and legislation. Stricter interpretation by the federal circuit court of the patent law requirements of utility and non-obviousness would limit private monopoly rights on basic research by reducing the proliferation of patents in the life sciences, including genetics research (Rai 1999, 137–40). A recent positive development is the federal circuit's *In Re Fisher* decision (421 F.3d 1365 [Fed. Cir. 2005]), holding that patent claims on expressed sequence tags (ESTs) encoding proteins and protein fragments in maize plants did not meet the patent law utility requirement.[42] While it is not clear whether this holding will be extended to ESTs in other cases (Cotropia 2006, 53), this case places an important limitation on patenting a basic research tool. The Patent and Trademark Office's (PTO) January 2001 adoption of stricter guidelines for application of the utility requirement is, similarly, a positive step.[43] Broader judicial interpretation of the common law experimental use exemption would also expand free access to patented inventions for noncommercial use, particularly in universities (Eisenberg 1989, 1017; Rai 1999, 139). Such reforms could also be accomplished through congressional amendments of the patent laws. The U.S. Supreme Court's decision not to review *LabCorp*, which concerned patentability of basic scientific knowledge, makes such congressional reform of patent law essential.

Federal agencies may adopt policies designed to broaden the public domain. The NIH has taken some steps in this direction, including its

decision not to file for patents on most research tools developed in its intramural research program (Rai 1999, 148; Symposium 2002, 390). The NIH
and other funding agencies may also issue grants with the condition of
placing the research results in the public domain, though the Bayh-Dole
Act limits federal agencies' ability to enforce this restriction on patenting
(Rai 1999, 148). In 1996, the National Human Genome Research Institute
and the NIH announced this condition for large-scale human sequencing
grants and, more recently, for grants for research on single nucleotide polymorphisms, or SNPs (147–48).[44] The NIH Guidelines issued in 1999 and
its 2004 "Best Practices for Licensing of Genomic Inventions" recommend
that NIH funding recipients limit their use of patents and exclusive licenses
on federally funded research tools and genomic inventions.[45] In February
2005 the NIH announced its policy that requests NIH-funded authors make
their articles available for free online access on its PubMed Central Web
site within twelve months of the official publication date (*NIH News* 2003).
Other laudable efforts, from both the public and private sectors, to expand
public access to research tools include the international Human Genome
Project's policy of placing data into the public domain and the more recent
nonprofit project Biological Innovation for Open Society, which uses licensing practices to broaden access to agricultural biotechnology research tools
(Kapczynski et al. 2005, 1069–72).

Legislative reform could strengthen the power of federal agencies to
require public dissemination of federally funded basic research. Congress
could amend the Bayh-Dole Act to expand a federal agency's ability to with-
hold title from government contractors or to require government contractors to issue nonexclusive licenses for use of their federally funded patents
(Rai 1999, 151). Congress could also amend the Bayh-Dole Act to give federal agencies the power to require government contractors to share basic
research tools with each other (113, 150–51). Arti Rai and Rebecca Eisenberg (2003, 56) have proposed amending the Bayh-Dole Act to give the
federal agencies the power to retain title to federally funded research beyond
the current legislative provision allowing federal agencies to restrict patenting in "exceptional circumstances."

Voluntary practices by universities and faculty may broaden dissemination and use of basic research discoveries. Given the widespread nature of
university involvement in such market activity, it is easy to forget that prior
to the Bayh-Dole Act, academic norms and practices disfavored patenting of academic research (Mowery et al. 2004, 42–43, 179).[46] In the context
of potentially lucrative biomedical research, there are some well-known

examples of faculty resistance to applying for patents, including Cesar Milstein and Georges Kohler's decision, in 1975, not to patent their Nobel Prize–winning invention of monoclonal antibody-producing hybridoma cells and their request to recipients of the cell line that they also not patent it (Rai 1999, 94; Eisenberg 1987, n.6, n.9; Kenney 1986, 129). Most universities, though, currently require faculty to sign agreements to disclose all their research inventions to the university and to assign all patent rights to the university (Rai 1999, n.180). Yet, even in this current commercial context of technology transfer, universities have adopted self-imposed restrictions on patenting. For example, Massachusetts Institute of Technology (MIT) avoids patents on basic research that is far from the point of practical commercial development, such as ESTs, but will engage in patenting and exclusive licensing of basic research where needed to encourage commercialization. Harvard and Stanford University do not patent ESTs and have a general policy not to patent SNPs (112). The growing number of reports and scholarly articles recommending against exclusive licensing may also encourage universities to broaden their use of nonexclusive licenses when they do engage in patenting (Pressman et al. 2006, 31–39).[47] Public interest organizations, such as Universities Allied for Essential Medicines, contribute to these voluntary efforts by advocating for universities' waivers of their patent rights and for use of nonexclusive licensing to make affordable generic medicines available in developing countries (Lieberwitz 2005, 149–50; Kapczynski et al. 2005).

Though the proposed reforms may have salutary effects, they do not address fundamental questions relating to the public interest. Policies that exclude some basic research from patenting and exclusive licensing may expand the public domain but do not challenge the premise that private entities should own and profit from publicly funded research through patents and licenses.[48] Nor do such policies challenge the premise that universities should be involved in market activities for the commercial interests of the university and industry. Thus, such reforms would not fully address the problems of restrictions on the public domain or the conflict between the university's commercial interests and the public interest.

The optimal means for reinvigorating the public domain and the university's public mission would be a full repeal of the Bayh-Dole Act, reinstating the pre-1980 policy of a presumption of government title to publicly funded research. The repeal of the Bayh-Dole Act would return public funding to its role of supporting research that is placed in the public domain. Given the central role of public funding in university research, this reform would also go a long way toward returning the university to its public mission to engage in research in the public interest. With the expanded public domain of federally funded inventions, legislators and federal agencies could turn their attention to other reforms to encourage wide use of such inventions, including measures to broadly publicize federally funded research results and to require federal agencies to adopt uniform licensing procedures. Such licensing regulations would not be needed at all where the federal agency does not patent a publicly funded invention, as the public may simply use the invention without a license.[49]

In addition to providing the "intangible" benefits of legitimacy of academic research, restoring federally funded research to the public domain will likely result in tangible public benefits, including lower costs of medical procedures or drugs. For example, the Tay-Sachs disease screening test, patented by the federal government, costs $100, while the privately patented screening test for two breast cancer genes costs $2,400 (Krimsky 1999, 37). Without the licensing fees, the cost for the genetic test for breast cancer is estimated at about $50 (Leroux 2001, C12).[50] Further, refocusing on the public domain may also have positive consequences for universities in the courts, as the courts may be more inclined to treat universities as unique research institutions that require special treatment such as a broad experimental use exemption from patent law.

Objections to repealing the Bayh-Dole Act will argue that businesses will not be interested in nonexclusive licensing of federally funded inventions, nor will businesses be interested in acting as federal contractors if their federally funded inventions will remain in the public domain. Such objections do not respond to the concern for using public funds in the public interest rather than as direct or indirect public subsidies for private business. The repeal of the Bayh-Dole Act would place business back in the position of maintaining ownership, including through patents, over research results in which they invest their own capital. Similarly, repealing the Bayh-Dole Act would avoid the indirect subsidy to industry resulting from university patenting and licensing of publicly funded research. Publicly funded university research would be available in the public domain through scholarly publications and through government title to resulting inventions. Objections to repealing Bayh-Dole do not take into account the record of industry's successful development of publicly funded research results placed in the public domain and accessible in scholarly publications and through nonexclusive licensing of government-owned patents (Eisenberg 1996, 1702–4; Lieberwitz 2005, 132–33). Further, as David Mowery and his coauthors note, industry's ability to develop products through exclusive licenses of university-owned patents does not answer the counterfactual of whether businesses could have developed these same products with broadly disseminated research results (Mowery et al. 2004, 183–84, 191; Johnston 2007, 163).

The same rationale supports reform of university policies in relation to privately funded research. Placing academic research into the public domain is most consistent with the university's institutional mission, eliminating the conflict between the public mission of the university and the private financial interests of the university or industry. Adopting a presumption against patents and exclusive licenses would revive policy followed by universities prior to the mid-1970s, including Harvard's policy that required approval from the president and fellows before obtaining university "patents primarily concerned with therapeutics or public health" and that such patents be taken only "for dedication to the public" (Eisenberg 1987, n.9). Businesses would still be free to contribute funds to the university in the belief that academic research findings will be of use to industry. Even if universities continue to apply for patents on privately funded research, the universities should accept private funding conditioned on agreement that any university-owned patents will be subject to nonexclusive licenses only to encourage broad dissemination of research results.

The proposed repeal of the Bayh-Dole Act may be criticized as being a pipedream—nothing more than a futile gesture under current economic and political conditions. Or, worse, it may be argued that advocating a full repeal of the Bayh-Dole Act and all university patenting distracts from the potential for successful adoption of more limited reforms. Any possible reforms, however, should be evaluated in terms of the goals the reforms seek to achieve. Thus, proposals short of repealing the Bayh-Dole Act are positive changes that should be measured against the ultimate goals of maintaining the communal values of scientific research and the public mission of the university. These goals are qualitatively different from a goal of commercializing academic research in a way that incorporates some public policy exceptions. The scope of the public domain for publicly funded research should not depend on the Supreme Court's holding in cases like *LabCorp*, the PTO's definition of the patent utility requirement, or amendment of Bayh-Dole to strengthen federal agencies' discretion to require nonexclusive licensing on a case-by-case basis. While the outcome in any of these judicial, administrative, or legislative changes would affect the scope of university patenting and licensing, only the repeal of the Bayh-Dole Act will restore the consistency between publicly funded research and the university's public mission. The rejection of patenting of publicly funded research would create a strong foundation for further university policy against patenting of privately funded research.

As the university acts for commercial interests, its identity changes from a public institution to a private market actor. By returning its priorities to fulfilling its public mission, the university can restore its legitimacy as an institution that serves the public interest. Reforms that strengthen the university's independence from commercial interests are essential to refocus academic research on the public domain. With this goal in mind, public policy and university practice can be shaped to benefit the public good.

Notes

1. For excellent discussions of the traditional norms of science, including Barber's and Merton's work, see Rebecca S. Eisenberg (1987, 181–84) and Arti K. Rai (1999, 88–94).

2. The university is described as a "public trust" in AAUP 1915 Declaration of Principles, in Louis Joughin (1967, 160). The university mission of contributing to the "common good" is discussed in AAUP 1940 Statement of Principles of Academic Freedom and Tenure, in *AAUP Policy Documents & Reports* (2001, 3). For examples of university mission statements describing goals of serving the public good, such as "expanding human knowledge" and benefiting society, see Josephine Johnston (2007, 164).

3. For how the biomedical industry relies particularly heavily on academic research and patents, see Josephine Johnston (2007, 161).

4. Rebecca S. Eisenberg (1987, n.1) defines "basic research" as "'pure' research directed solely toward expanding human knowledge, as opposed to 'applied' research directed toward solving practical problems," and notes that "whatever the validity this dichotomy may have in other contexts, it is difficult to maintain in the context of contemporary biotechnology research."

5. NIH funding comprised 60.5 percent of federal funds for U.S. academic research in 2001 and 66 percent in 2003. See Mowery et al. (2004, 25) and Johnston (2007, 161); both sources cite National Science Board statistics.

6. Patent & Trademark Act Amendments, 35 U.S.C. secs. 200–212, commonly known as the Bayh-Dole Act. Where the federal funding recipient chooses to apply for a patent, the federal funding agency retains a nonexclusive license to use the publicly funded invention. See Rebecca S. Eisenberg (1987, 196).

7. The Bayh-Dole Act was amended to expand patenting rights from small businesses to all businesses receiving federal research funds, regardless of the size of the business (Eisenberg 1996, n.180).

8. See Rai and Eisenberg (2003, 54). "Concern over U.S. economic competitiveness" contributed to congressional support for the passage of the Bayh-Dole Act (David C. Mowery et al. 2004, 181).

9. One biotechnology company, Incyte, has reportedly claimed that it has filed 50,000 patent applications on pieces of various genes (Symposium 2002, 387).

10. Mowery et al. (2004, 180–83) suggest that the Bayh-Dole Act did not cause the dramatic increase in university patenting activity but, rather, had a primarily quantitative effect of accelerating existing trends in universities toward increased patenting. Given the exponential growth in patents after Bayh-Dole, however, it seems accurate to characterize this change in university patenting policy and practice as qualitative in nature. Indeed, Mowery et al. acknowledge the role of Bayh-Dole in increasing the patenting at universities that had already been involved in such activities as well as bringing in universities that had not yet been involved in patenting.

11. J. Johnston (2007, 162) cites National Science Board statistics.

12. Overall, industry funding of science and engineering research and development in universities decreased during FY 2002–4. In FY 2004, federal funding of such research was $27.4 billion, an increase of 10.7 percent from the previous year, and industry funding was $2.1 billion, falling by 2.6 percent from the previous year (National Science Foundation 2006).

13. Other examples of strategic corporate alliances include: the 1974 grant of a twelve-year $23 million grant from Monsanto to Harvard Medical School in exchange for Monsanto's right to a worldwide exclusive license for resulting inventions; the 1982 Monsanto agreement with Washington University for $23.5 million over five years in exchange for exclusive licensing rights to resulting biomedical patents (after being renewed three times, Monsanto's financing of the university came to about $100 million.); the 1994 MIT agreement with Amgen for $30 million of corporate funding over ten years to the Departments of Biology and of Brain and Cognitive Sciences in exchange for patents to be jointly held by MIT and Amgen. See Kenney (1986, 58–60, 67–69); Krimsky (1999, 28–29); Lawler (2003, 330); "MIT's Alliances with Industry," *MIT News*, http://web.mit.edu/newsoffice/nr/2000/alliance.html (accessed 22 Jan. 2007).

14. J. Johnston (2007, 162–63) cites 2004 statistics from the Association of University Technology Managers.

15. Dueker (1997, 478–79) cites the $300 million in gross royalties to universities from licensed technologies in FY 1995, which was equivalent to 2.6 percent of federal funding to universities. In FY 2004 gross royalties to universities from such licenses was $1.39 billion, which was equivalent to 5 percent of the $27.4 billion of federal funding to university research in the United States. See Association of University Technology Managers (2005); National Science Foundation (2006).

16. See, for example, Bok (2003, 141) for the conclusion that such commercial activities promote the public interest; and Mowery et al. (2004, 90–91) for a description of universities' lobbying activities in support of the Bayh-Dole Act.

17. See Mowery et al.'s (2004, 190) criticism of "a single-minded focus on patenting and licensing as the only important or effective channels for technology transfer."

18. See Mowery et al. (2004, 191) for a recommendation that "universities should pursue nonexclusive licenses for the fruits of publicly funded research whenever possible."

19. For examples of publicly funded academic research that resulted in university patents that were exclusively licensed to industry and marketed as high priced drugs or medical tests, see Lieberwitz (2005, 127–28, 133–34).

20. On the relationship between university independence and its institutional legitimacy, see Slaughter and Leslie (1997) and Henry Etzkowitz et al. (1998).

21. Mowery et al. (2004, 185–86) describes results of a 1993 survey by Blumenthal et al. (1997).

22. Press and Washburn (2000, 42) describe a case in 1996 where four university science researchers resigned after their corporate sponsor, Sandoz (now Novartis), altered a manuscript to remove findings of potential negative effects of the corporation's drugs.

23. The journals will include notice of an author's refusal to disclose this information (Huvane 2002, 7).

24. Petersen (2002, A1) describes a "1998 survey of named authors writing for some of the nation's top journals, including *The Journal of the American Medical Association,* which published the survey, . . . that 11 percent of the articles had been ghostwritten."

25. Former Harvard University president Derek Bok (2003, 152–53) concludes that there is no current need for reform to limit such large-scale corporate funding arrangements because they are unlikely to proliferate.

26. Mowery et al. (2004, 185–86) describes results of a 1993 survey by Blumenthal et al. (1997) and by Campbell et al. (2002).

27. For examples of university and industry claims to rights of reach through patents, see Kleinman (2003, 123–25) for DuPont's requirement of royalty payments on commercial discoveries using its patent on genetically altered Cre-*lox*P mice; Rai and Eisenberg (2003, 57) on the patent infringement suit by patent owners Harvard, MIT, and the Whitehead Institute for Biomedical Research and exclusive licensee Ariad Pharmaceuticals against Eli Lilly & Co. claiming the right to royalties on sales of products developed with the use of the patented research tool, nuclear factor kappa B.

28. The oncomouse is genetically engineered to be predisposed to cancer (Murray 2006, 1).

29. Murray (2006, 25–31) describes the campaign of resistance by academics to DuPont's terms, which resulted in more open access to oncomice in academic research.

30. *Metabolite Labs., Inc. v. Lab. Corp. of Am. Holdings,* 370 F.3d 1354 (Fed. Cir.

2004), cert. granted in part, 126 S. Ct. 601 (2005), cert. dismissed as improvidently granted, 2006 U.S. LEXIS 4893.

31. See *Diamond v. Diehr*, 450 U.S. 175, 185 (1981): "Excluded from . . . patent protection are laws of nature, natural phenomena, and abstract ideas."

32. Brief for Respondents, *Laboratory Corporation v. Metabolite Laboratories* (2006, 8), hereafter LCML; see also Pollack (2006, C1).

33. 2006 LEXIS 4893, p. 15.

34. 2006 LEXIS 4893, pp. 2, 13. The Court dismissed the writ of certiorari in a per curiam opinion without further explanation. As Justice Breyer explained, the Court may have decided to dismiss the case on the ground the issue of patentability of a natural phenomenon was not properly preserved by the defendant in the lower courts or that the federal circuit court had not directly considered the issue.

35. 2006 LEXIS 4893, p. 23.

36. 2006 LEXIS 4893, p. 4.

37. "High Court Hears Patent Arguments" (2006, C2).

38. See also Pollack (2006, C1).

39. *Roche Prods, Inc., v. Bolar Pharmaceuticals Co.*, 733 F.2d 858, 863 (Fed. Cir.), cert. denied, 469 U.S. 856 (1984).

40. The statutory "safe harbor" exemption was created in 1984 to support research related to generic drug development (35 U.S.C. 271[e] [2000]).

41. Garde (2005, 265–66) notes that the Court declined to comment on the application of the statutory research exemption to research tools, which were not at issue in the case.

42. Expressed sequence tags (ESTs) are "gene fragments, typically of unknown function, which will typically be used as probes to locate and characterize the full gene" (Rai 1999, 104).

43. For a helpful chart see Patent and Trademark Office, "Utility Examination Guidelines," 66 *Fed Reg.* 1092 (Jan. 5, 2001), http://www.uspto.gov/web/offices/pac/dapp/ogsheet.html (accessed Jan. 22, 2007).

44. Mikhail (2000, nn.38, 392) cites National Institutes of Health, "Report of the National Institutes of Health Working Group on Research Tools" (1998), http://www.nih.gov/news/researchtools/index.htm. Rai (1999, 105) explains that single nucleotide polymorphisms (SNPs) "represent those areas in which human genomes differ by only one DNA base pair from one another. It is estimated that there are approximately 300,000 SNPs in the human genome."

45. See J. Johnston (2007, 165–68); Garde (2005, 272–73).

46. Kenney (1986, 32) describes the "ideology . . . of scientists working for the public good" as leading to the view that "industry's motives—especially that of profitability—were suspect, and the applied science orientation of industry . . . to be scientifically uninspiring to scientists."

47. J. Johnston (2007, 166–68) discusses the range of available patent and licensing approaches for universities.

48. See Krimsky (1999, 38).

49. A federal agency may decide that it is in the public interest to apply for a patent on federally funded inventions to ensure that a private party does not apply for a patent and to avoid litigation over the issue of prior publication of the invention (Kenney 1986, 111–13; Eisenberg 1988, 1377).

50. The discoveries of the breast cancer genes were based, in part, on NIH-funded academic research (Williams-Jones 2002, 132–33). As Daniel Kleinman (2003, 131–32) reports, Myriad Genetics asserts that it would not have invested the research and development funds related to the screening test without patent protections.

4

Patterns of Research and Licensing Activity of Science and Engineering Faculty

JERRY G. THURSBY AND MARIE C. THURSBY

INTRODUCTION

The importance of university research for industrial innovation is widely accepted, so much so that any changes in the research environment tend to spark controversy. The recent increase in university licensing is no exception. The eighty-four universities responding to the Association of University Technology Managers (AUTM) Survey in 1991 and 2000 reported that over the period invention disclosures increased 84 percent, new patent applications increased 238 percent, licenses executed increased 161 percent, and royalties increased 520 percent. While technology managers and university administrators cite such figures as evidence of the increasing contribution of universities to the economy, skeptics question the impact of licensing on the conduct of university research.

Central to the debate is faculty behavior. Proponents of licensing argue that without the incentives it provides, neither faculty nor companies would undertake the development needed for many results of federally funded research to be transferred to industry. This, of course, is the key premise underlying the Bayh-Dole Act, which since 1980 has allowed universities to own and license results of federally funded research. Critics of the act claim that academic publication would be sufficient for industry to pick up university inventions, and more importantly, as the chapter by Lieberwitz demonstrates, they express concerns that potential financial returns from licensing have diverted faculty from more basic to applied research.

There are three major issues related to faculty behavior:

1. Does the increase in licensing reflect a shift away from fundamental research or simply an increased willingness of faculty to have their

work licensed as well as published (or both)? While the latter was intended by Bayh-Dole, the former was not and has been questioned in recent congressional hearings as well as by the National Academies Committee on Science, Engineering, and Public Policy (COSEPUP).

2. How is the propensity of faculty to disclose inventions related to individual characteristics (e.g., publication record, research sponsorship) and other characteristics (e.g., academic discipline, university and/or department effects)?

3. What is the life-cycle behavior of faculty with respect to research, publication, and license-related activity? Evidence on life-cycle productivity will contribute to a better understanding of the science and technology enterprise in general, as the age distribution of scientists and engineers changes (National Science Board 2000).

Little is known about these issues, largely because the literature on university licensing has focused on disclosures, patents, and licenses aggregated by field or university rather than individual inventors. We do not know even the portion of faculty who engage in licensing, much less their personal and research characteristics as compared to faculty who do not engage in licensing. In this study we seek to redress this problem by examining the personal and research characteristics of a sample of science and engineering faculty at six major universities.

The database we examine has the publication, disclosure,[1] and personal profiles of 3,241 faculty members at Cornell University, Massachusetts Institute of Technology (MIT), University of Pennsylvania, Purdue University, Texas A&M University, and University of Wisconsin–Madison over a seventeen-year period. These data allow us to examine the evolution of faculty research and licensing behavior and the extent to which it is related to individual characteristics such as age or to environmental factors such as academic quality of their department. Further, we can address differences in patterns across universities and across disciplines in the life and physical sciences as well as engineering.

What Do We Know about Licensing and Faculty Behavior?

There is growing evidence suggesting that, for many inventions, technology transfer would be difficult in the absence of incentives provided by patent licensing. Recent surveys show that the overwhelming majority of university inventions are so embryonic that commercial application requires not only further development but also faculty cooperation in that development (Thursby et al. 2001; Agrawal and Henderson 2002; Thursby and Thursby 2002). Jensen and Thursby (2001) show that, to the extent faculty prefer research to development, financial incentives are needed to induce them to work with licensees in further development. The emerging picture is that while some inventions, particularly those needing little development or faculty involvement (see Colyvas et al. 2002), would be transferred in the absence of licensing, the most embryonic might not.[2]

We know little, however, about the effect of licensing on faculty research as the literature on industry-university collaboration and research relates primarily to consulting and sponsored research, and the evidence is mixed. While some studies suggest that applied research has increased recently (Cohen et al. 1994; Morgan et al. 1997; Rahm 1994), others point to a long history of such research (Mowery and Ziedonis 2002; Mowery et al. 1999; Rosenberg and Nelson 1994). Cohen et al.'s (1994) survey of university-industry research centers (UIRCs) provides evidence of the countervailing effects of industry collaboration on faculty productivity, with so-called commercial outputs of research increasing and publications decreasing (except in biotechnology). Given the importance of publications for industrial productivity (Adams 1990), these results are cause for concern. By contrast, Mansfield (1995), Zucker et al. (1998b, 1999), Stephan et al. (forthcoming), and Murray (2002) find a complementary relationship between research productivity and commercial activity. Mansfield's (1995) study of 321 academic researchers found that faculty frequently worked on basic problems suggested by their industrial consulting. Similarly, Zucker et al. (1998b, 1999) found that the most productive scientists in biotechnology often start new enterprises while continuing research in their academic appointments. In the case of tissue engineering, Murray (2002) shows that many results are both patented and published.

Stephan et al. (forthcoming) comes the closest to examining our questions of interest. Using individual level data from the 1995 Survey of Doctorate Recipients, they examine the relationship of patents and publications as well as life-cycle and other individual effects. They find a complementary relationship for patents and publications. They also find that nontenured faculty are more likely to patent than are tenured faculty. For our purposes, the cross-sectional nature of these data presents a drawback. Further, while patents clearly signal faculty willingness to engage in licensing activity, the use of patents excludes information about faculty whose inventions are licensed but not patentable. Their approach also excludes observations on faculty who show an interest in licensing by disclosing inventions but whose inventions are not deemed commercially viable (and as a result are not patented).

Only a few studies examine invention disclosures and faculty research. Lach and Schankerman (2002) find a positive relationship between invention disclosures and the share of license revenue accruing to inventors. While they interpret this as showing the responsiveness of research to financial incentives associated with licensing, we argue that disclosures show the faculty's willingness to engage in licensing and may or may not reflect changes in research agendas. In Thursby and Thursby (2002) we examine whether the growth in university licensing is driven by an unobservable change in the propensity of faculty and administrators to engage in license activity. We find that changes in the direction of research are relatively less important than increases in the propensity of administrators to license inventions and in business reliance on external research and development (R&D). However, our data are not at the level of the individual scientist but, rather, research outputs at the university level.

To our knowledge, there are no empirical studies of the relationship between licensing activity and research productivity at the level of the scientist. However, recent theoretical work has examined the faculty member's choice when his/her income is a function of license revenue as well as his/her university salary. Jensen and Thursby (2004) examine a model in which faculty choose the amount of time to spend on basic and applied research given their salary and the share of license income they receive. In their model, faculty enjoy both the puzzle-solving aspects of research (see Hagstrom 1965 and Stern 2004) as well as the prestige from successful research (see Stephan 1996). They show that with these effects factoring into the scientist's decision, the opportunity to earn license income may well not change his/her research agenda. Finally, Thursby et al. (2005) examine a model in which faculty incentives to conduct basic and applied research change over the life cycle. In their model, basic research is publishable while applied can be licensed. As the faculty approach retirement, the incentive for publishable research declines relative to applied since the faculty can collect their share of license revenue beyond retirement.

THE DATA

We examine the research, demographic, and disclosure profile of a group of faculty scientists and engineers at six major universities. Note that our choice of universities is not random. Given our interest in the effect of licensing on faculty research, it is important to select major research universities with substantial licensing activity. As shown in Table 4.1, all of the universities in the sample are among the top fifty universities in terms of total research expenditures, licenses executed, and invention disclosures as reported in the 2001 AUTM Survey. All but one of the universities have licensing activity above the average of the top fifty licensing universities.

Our measure of faculty interest in licensing is invention disclosures rather than licenses executed. While disclosures and licenses are not independent, we believe the former is more representative of faculty interest and the latter more representative of commercial quality. That is, a license disclosure simply indicates that an inventor has a research result that he/she believes has commercial potential. While all universities in the sample require their employees to file such disclosures, this is hardly enforceable. Faculty may not disclose for a variety of reasons. In some cases they may not realize the commercial potential of their ideas; but often faculty do not disclose inventions because they are unwilling to risk delaying publication during the patent and license process.[3] Faculty who specialize in basic research may not disclose because they are unwilling to spend time on the applied research and development that is often needed for businesses to be interested in licensing university inventions (Thursby and Thursby 2002; Jensen et al. 2003). While a disclosure signals a willingness to be involved with licensing, it need not indicate that the research was motivated by the desire to license. Curiosity-driven research can often lead to commercially applicable results by accident. In their interviews with MIT mechanical engineering faculty, Agrawal and Henderson (2002) found that most conducted research with the primary goal of publishing.

TABLE 4.1 Research and licensing activity in 2000.

University	Total research expenditures	Licenses executed	Invention disclosures
MIT	$727,600,000	102	425
Univ. of Wisc.–Madison	$554,361,000	127	277
Univ. of Penn.	$529,554,951	63	223
Texas A&M	$397,268,000	58	140
Cornell	$396,900,000	63	177
Purdue	$263,440,339	76	118
Average of top 50 universities	**$325,195,723**	**51**	**137**

The faculty we study are from the list of science and engineering faculty in Ph.D.-granting departments given in the 1995 National Research Council (NRC) report. We exclude faculty not listed in such departments, which means medical school faculty are excluded unless they also hold appointments in Ph.D.-granting departments. Departments also are excluded if one could not reasonably expect disclosure activity (as in, for example, astronomy departments).

The technology transfer office (TTO) of each university supplied us with the names of disclosing faculty as well as dates of disclosure. Four universities provided disclosure information for 1983 to 1999, and the others provided information from 1983 to 1996 and from 1987 to 1999.[4] Matching these files with the NRC list provides a sample composed of multiple years of disclosure (or nondisclosure) activity for faculty of our universities in 1992. We have information on dates of hire and departure (if applicable) so that the final sample includes the faculty when they were actually at the respective universities.[5] In our sample we have 3,342 faculty and 44,731 observations where an observation consists of a person/year.

Thus, for each faculty member in our sample, we know whether or not he/she disclosed (and, if so, how often) in each year that he/she was on the faculty of his/her respective university during the period made available by his/her TTO. Of the 44,731 observations, 3,241 (7.1 percent) represent disclosures in a particular year by a faculty member. This is our measure of faculty *interest* in licensing activity. In the remainder of the study, we use disclosure to indicate that a faculty member has disclosed at least once in a given year.

Given the concern that academics have become too commercial, the portion of faculty expressing interest in licensing is remarkably low. Of the 3,342 faculty, 2,145 (64.2 percent) never disclosed an invention, 495 (14.8 percent) disclosed in only one year, and 254 (7.6 percent) disclosed in only two of the years they were included in the sample. Only sixty-seven faculty (2.0 percent) disclosed in eight or more of the years they were in the sample. Across the six universities, the fraction of faculty who never disclosed ranges from 53.9 percent to 72.2 percent. This, of course, does not tell us which faculty members disclosed: for example, was it the most productive in terms of publication? More to the point, simple counts reveal nothing about changes in the nature of the research conducted by the 35.8 percent who disclosed.

DISCLOSURE ACTIVITY AND FACULTY CHARACTERISTICS

In this section we present simple tabulations of disclosures and researcher characteristics. Given the number of observations it is common to find statistically significant correlations between disclosures and other variables, however small that relationship might be. Thus, it is not surprising that in every case we reject the null hypothesis that the tabulations are random.

Table 4.2 gives the distribution of observations and disclosures by university. For purposes of confidentiality, we provide only the distribution across universities and do not identify the university (they are *not* listed alphabetically). The second column is the percentage of observations from a particular university that are disclosure observations. This is the probability (stated in percentages) that an observation from some university is a disclosure observation, that is, the probability of disclosure conditional on university, or PR (Disclosure | Univ.). For example, in the first row we know that 4.41 percent of the observations from this university are disclosures.

Disclosure activity varies substantially across the six universities. Less than 5 percent of all observations from Universities 1 and 5 are disclosures, whereas more than 13 percent of University 2's observations are disclosures. An interesting question, and one we cannot answer here, is how this dispersion relates to university policies and culture.

Table 4.3 gives the distribution of observations across the years of our sample. The table shows a dramatic increase in disclosure activity from 1983 until the mid-1990s, at which time activity leveled off with 10–11 percent of a year's observations being disclosures. In 1983 the probability that a faculty member disclosed was only 1.11 percent as compared with more than 10 percent by 1996. Earlier we noted that only about 20 percent of the faculty disclose in more than one year in our sample. Here we find that, in the later years, about one in ten faculty are disclosing in a given year.

TABLE 4.2 Distribution of observations and disclosures.

University	PR (Disclose|Univ.)
1	4.41
2	13.82
3	9.27
4	6.25
5	4.91
6	6.81

Thus, while disclosure activity has increased, it tends to be concentrated among a few faculty. Finally, the last column in Table 4.3 gives publications per faculty member for the years reported. Contrary to the notion that disclosures may come at the expense of (or accompany a decline in) publications, publications per faculty in our sample more than doubled. Assuming that a publication in 1983 reflects the same research productivity as it does in 1999, the increased disclosure activity may in fact reflect increased research activity. Of course, as noted earlier, publication counts tell us nothing about the nature of research. If, as feared, research has become more applied, it may well be the case that applied research, in general, leads to higher numbers of publications for the same research effort.

To examine the nature of research, we map each faculty member's journal publications into Narin et al.'s (1976) classification of the "basicness" of journals. This classification characterizes journals by their influence on other research. As discussed by Narin et al. (1976), basic journals are cited more by applied journals than vice versa, so journals are considered "basic" if they tend to be heavily cited by other journals. For example, if journal A is heavily cited by journal B, but B does not tend to be cited by A, then A is said to be a more basic journal than is B. Advantages of the Narin classification are not only its measure of influence but also its ease of extending

TABLE 4.3 Disclosure distribution by year.

Year	Percent of sample	PR (Disclose|Year)	Publications per faculty
1983	3.64	1.11	1.62
1984	3.73	1.50	1.64
1985	3.84	2.16	1.70
1986	3.98	2.59	2.49
1987	6.08	5.01	2.36
1988	6.37	6.12	2.48
1989	6.61	6.24	2.33
1990	6.88	5.64	2.35
1991	7.15	7.74	2.42
1992	7.30	8.16	2.57
1993	7.16	9.08	2.71
1994	7.10	9.31	2.85
1995	6.86	9.44	2.92
1996	6.75	10.40	3.11
1997	5.58	11.45	3.61
1998	5.53	11.35	3.84
1999	5.45	10.27	3.69

the measure to a large number of journals and articles. The ratings are on a five-point scale, and we classify as basic only publications in the top basic category, which covers about 62 percent of all ranked journal publications. Only about a third of the publications could be rated, but we found no systematic change over time in the number of publications that could be rated. In a regression of the fraction of rated publications (where we drop observations with no publications, rated or otherwise) on a set of indicator variables for the year of the observation, we found an R^2 of only 0.0016 and very few significance differences in the coefficients of early versus later years.

We combine our information on annual faculty activity with information on age, year of Ph.D., academic field, gender, and quality of the faculty member's department. In many cases, birth dates are unavailable; in such cases we assume birth dates are twenty-one years prior to year of undergraduate degree; or, if date of undergraduate degree is not available, we assume birth year was twenty-nine years prior to the date of Ph.D. The department quality measure is taken from the National Research Council's (1995) survey. Departments are rated on a six-point scale from 0 to 5 where 5 is distinguished. Table 4.4 gives summary statistics for the data. For gender, we present the percent of the sample that is male.

Tables 4.5 and 4.6 provide information on publications and the portion that are basic in relation to disclosures. Not surprisingly, those who publish more are more likely to disclose. What is somewhat surprising is that 5 percent of the time when a faculty member had zero publications he/she still disclosed an invention. Note, however, we are considering only contemporaneous publications. Nonetheless, there are some faculty who disclose at some point but have zero publications for all the years they are in the sample. Those who disclose at least once in our sample account for 42.9 percent of all publications.

TABLE 4.4 Summary statistics.

Variable	Mean	Standard deviation
Age	48.35	10.91
Year of Ph.D.	1970.76	10.46
Quality	3.91	0.60
Publications	2.67	8.17
Percent male	91.3	

In Table 4.6 we drop faculty who did not have any publications in journals rated by Narin (1976); this leaves a sample of 11,577 observations. We
then tabulated the fraction of basic publications with disclosure activity.
The probability of disclosure initially increases with the fraction of basic
publications and then decreases. Those whose research is in the midrange
(33 percent to 67 percent) have the highest probability of disclosure (18.9
percent). In the introduction we noted an interest in the behavior of basic
research not only with respect to disclosure activity but also with respect
to time. To examine this question, we continue with the reduced sample of
11,577. For this group of observations we assume that the fraction of total
publications that are basic is the same as the fraction of rated publications
that are basic. The number of basic publications per faculty member by
year is then computed and presented in the second column of Table 4.7.
In the third column we present the number of total publications per faculty member by year. While these averages have varied over the seventeen
years, the two columns are very closely related. The simple correlation
between these two columns is 0.95. The implication is that there has been
little or no change in the relation between total publications and the percent that are basic publications.

Table 4.8 relates gender to disclosure activity. There is a substantial disparity between men and women. Only 5.54 percent of the observations for

TABLE 4.5 Annual publications.

Pubs.	Percent of sample	PR(Disclose|Pubs.)
0	49.67	5.01
1–2	23.85	6.41
3–4	11.76	9.70
5–6	5.60	11.09
>6	9.12	15.28

TABLE 4.6 Percent of basic research.

Percent	Percent of sample	PR(Disclose|Basic)
No Basic	26.14	11.34
<33	1.26	15.65
33–67	7.62	18.90
67–100	64.98	8.92

women are disclosure year observations, whereas 7.61 percent of the observations for men are disclosures.

Tabulations of age and disclosure activity are in Table 4.9. Disclosure activity tends to rise until the middle age of 45–55, after which it declines. This pattern, of course, may reflect the fact that publications (and hence research) tend to follow that same pattern. In the last column of Table 4.9, we present the percentages of those in the different age categories who have at least one publication in a given year. Note that, indeed, publications first rise and then fall (after ages 35–45). There is one final point about age and our sample. Since our sample includes all faculty members in certain departments in 1992, the average age of faculty in the sample is rising. In 1983 the average faculty in our sample was 43.4 years, while in 1999 it was 55.2 years.

TABLE 4.7 Basic publications by year.

Year	Basic publications/faculty	Total publications/faculty	Ratio of basic to total [column 2 divided by column 3]
1983	6.41	7.24	0.885
1984	5.79	6.94	0.834
1985	5.47	6.59	0.831
1986	5.08	6.43	0.791
1987	4.43	6.10	0.725
1988	4.56	6.14	0.742
1989	4.26	5.91	0.720
1990	4.18	5.96	0.700
1991	3.94	5.66	0.696
1992	4.56	6.25	0.729
1993	4.75	6.43	0.738
1994	5.09	6.69	0.761
1995	5.21	6.92	0.754
1996	5.61	7.45	0.753
1997	6.33	8.02	0.789
1998	6.95	8.62	0.806
1999	6.74	8.70	0.775

TABLE 4.8 Gender and disclosures.

	Percent of sample	PR(Disclose|Gender)
Male	91.3	5.54
Female	8.7	7.61

For each faculty member, we have information not only on age but also on year they received their Ph.D. It is possible that there are Ph.D. "cohort" effects not captured by age. That is, while age and year of Ph.D. are highly correlated in this sample (the simple correlation is -0.87), there is likely to be independent information regarding year of Ph.D. that is not captured by age. For example, for a faculty member in the sample over the years 1983–99, age varies but year of Ph.D. does not. If year of Ph.D. reflects to some extent the knowledge base of the researcher, and if the likelihood of disclosure is affected by the state of the field at the time of disclosure, then the cohort effect may well affect the likelihood of disclosure. This effect will not be captured entirely by age even though age and year of degree

TABLE 4.9 Age and disclosures.

Age	Percent of sample	PR(Disclose|Age)	Percent with publications
<35	11.88	6.88	39.70
35–45	30.46	7.79	54.26
45–55	30.48	8.06	53.83
55–65	20.4	6.20	51.19
>65	6.79	6.42	49.36

TABLE 4.10 Year of Ph.D.

Area	Year	Percent of sample	PR(Disclose|Cohort)
Biological sciences	<1960	16.84	5.56
	1960–1970	27.57	6.27
	1970–1980	37.41	8.90
	1980–1990	17.77	8.47
	1990–1999	0.42	10.45
Engineering	<1960	15.55	7.88
	1960–1970	30.38	7.81
	1970–1980	29.34	9.55
	1980–1990	23.10	11.99
	1990–1999	1.64	14.52
Physical sciences	<1960	20.43	3.56
	1960–1970	30.89	3.61
	1970–1980	28.81	5.08
	1980–1990	19.17	6.34
	1990–1999	0.70	0.00

are correlated. We use the NRC's general field classification to break the sample into the three major program areas: biological sciences, engineering, and physical sciences. In Table 4.10 we tabulate disclosure activity with Ph.D. cohorts where we use the decade of Ph.D. to classify the cohort. For biological and physical sciences the faculty who completed degrees in the '70s are most likely to disclose. For engineering it is the latest cohorts.

For each faculty member, we have department characteristics from the NRC survey. The characteristics of particular interest here are the major program area of the researcher and the academic quality of their department.[6] It is expected that disclosure activity will vary according to the major program field of faculty. This follows both from the nature of research (for example, engineering is more applied than physical science) and the market demand for technologies (for example, biological science results are more in demand by industry[7]). Table 4.11 provides a breakdown by these three program areas. Engineering is the most active, and biological sciences is more active than physical sciences.

In Table 4.12 we consider quality of departments and likelihood of disclosure. Our department academic quality rankings come from the NRC's (1995) academic quality survey. In that ranking a score of 5 is distinguished. As the universities considered here are all major research universities, it is not surprising that less than 10 percent of their science and engineering faculty are in departments ranked below 3. However, it is these faculty that are most likely to disclose.

TABLE 4.11 Distribution by field.

Area	Percent of sample	PR(Disclose|Area)
Biological sciences	37.14	7.73
Engineering	33.98	9.47
Physical sciences	28.88	4.58

TABLE 4.12 Distribution by quality.

Quality	Percent of sample	PR(Disclose|Area)
< 3	9.53	11.90
3–4	46.64	5.80
4–5	43.83	8.15

Note: 5 is distinguished.

A Logit Model of Disclosure

A weakness in what we have done above is that, with few exceptions, we have looked at bivariate relationships between disclosure activity and other variables of interest. We have not considered disclosure activity relative to some variable *holding constant* the values of other variables. Here we use a probability model of disclosure wherein the probability of disclosure is modeled as a linear function of observables. This is purely a descriptive exercise in that we do not have a formal model of disclosure activity, nor do we account for potential econometric problems that a formal model might suggest.[8] Future research will consider such a model of faculty behavior and what it implies about the proper specification of an econometric model.

In Table 4.13 we present the logit results. (The reader can find a brief description of the logit technique in the statistical methods section of the introduction of this volume.) Disclosures are coded as ones, and results are presented as odds ratios. An odds ratio gives the effect of a unit change in a right-hand-side variable on the ratio of the probability of a disclosure divided by one minus the probability of a disclosure. Hence, an odds ratio of less than one implies that the right-hand-side variable has a *negative* effect on the probability of disclosure. After dropping observations with missing data, the number of observations is 40,869. The pseudo R^2 is 0.0839, and the chi-square statistic for overall fit has a p value of less than 0.0000.

Age and a more recent year for the Ph.D. have negative effects on disclosure. For example, a forty-year-old who received his/her degree in 1980 is less likely to disclose than a forty-year-old who received his degree in 1970. However, over time, any given individual is less likely to disclose. Thus, as time passes, a faculty member is less likely to disclose because he/she is getting older and further from the year of Ph.D. Department quality also has a negative effect on the probability of disclosure. Not surprisingly, the likelihood of disclosure increases as publications increase. What is dramatic from this regression is the greater likelihood of a male disclosing an invention. Holding constant a number of other factors, a male faculty member is about 62 percent more likely to disclose in any given year.

With regard to field of research, those in biological sciences and engineering are equally likely to disclose, all else constant, whereas both are more likely to disclose than are faculty in the physical sciences. Across the six universities there are significant differences in the probability of disclosure, and we note some striking differences in probabilities of disclosure across these universities.

TABLE 4.13 Logistic model.

Indicator	Variable	Odds ratio	t-Stat	Significance
Personal	Age	0.94	−6.46	*
	Male	1.62	6.05	*
	Publications	1.02	9.66	*
	Department quality	0.91	−5.76	*
	Year of Ph.D.	0.96	−4.54	*
Year	1984	1.56	1.28	
	1985	2.14	2.29	†
	1986	2.98	3.46	*
	1987	7.14	6.83	*
	1988	9.06	7.69	*
	1989	10.17	8.07	*
	1990	9.69	7.82	*
	1991	13.86	9.07	*
	1992	15.21	9.31	*
	1993	18.25	9.84	*
	1994	20.37	10.10	*
	1995	22.42	10.29	*
	1996	25.48	10.58	*
	1997	31.83	11.12	*
	1998	32.72	11.04	*
	1999	32.02	10.79	*
Field	Engineering	1.09	1.88	
	Phyical sciences	0.52	−11.80	*
University	UNIV-2	3.07	16.09	*
	UNIV-3	1.69	7.34	*
	UNIV-4	1.42	4.34	*
	UNIV-5	0.77	−3.11	*
	UNIV-6	1.24	2.98	*

* Significant at 1% level
† Significant at 5% level

Regarding year effects, we find in the bivariate comparison of year and disclosure that only about 1 percent of faculty disclosed in 1983 whereas more than 10 percent disclosed by 1996. Holding constant other effects on disclosure activity, we find much the same effects. Tests of equality of regression coefficients reveal that the coefficients of the 1997 through 1999 indicator variables are not significantly different from one another, but the 1997 through 1999 coefficients are significantly different from all other year effects, and the 1996 coefficient is significantly different from all prior years with the exception of 1995. Essentially, we find results in our logit regression that are similar to the results in Table 4.3. The probability of disclosure rises from 1983 until around 1996, when the annual disclosure rate tends to level off.

CONCLUDING REMARKS

A few years ago, an *Atlantic Monthly* cover story titled "The Kept University" questioned whether the academic enterprise as we know it has suffered with the increased financial incentives for faculty to engage in commercial activities such as licensing (Washburn 2000). Despite the publicity and obvious importance of the underlying issues, we know relatively little about the relationship of such activity to faculty research, in large part because the data needed to characterize the relationship are at the level of the individual faculty member. Most of what we know comes from data provided by university TTOs (either through AUTM or other research from TTO files or surveys). With this project, we attempt to reduce the gap in our understanding by building a database that allows us not only to examine the behavior of faculty engaged in licensing but also to compare these faculty with those who have avoided the process. Here we provide a first analysis of these data.

Our analysis is preliminary and should be viewed with caution, as we have avoided any formal economic and econometric modeling. Moreover, the sample represents faculty from six universities. These caveats aside, the results provide a striking picture. Note that while the universities in our sample are above the average of the top fifty universities in terms of both research and licensing activity, we find that only a small portion of faculty engage in licensing. Indeed, 80 percent of the faculty in our sample either never disclosed or disclosed only once in the seventeen-year period. However, we found a dramatic increase in the portion of faculty who became involved over the period. The percent of faculty disclosing in a given year increased from 1 in 100 in 1983 to 1 in 10 by 1996, though it seems to have stabilized for the years 1997–99. So the much-publicized increase in licensing activity appears to be concentrated among a minority of faculty. Moreover, this increased licensing activity does not appear to signal a change in the direction of faculty research for our sample. The portion of publication that is basic by Narin's citation-based index was relatively constant over the period.

We find significant results regarding age, gender, publications, department quality, and year of Ph.D. What we find a bit surprising is that more recent Ph.D.s are less likely to disclose (controlling for age). Recall, however, that these results are based on a purely descriptive model. Results such as this beg for a structural model that would allow us to control for other factors, such as funding and university policy patterns.

Notes

Financial support was provided by the Alan and Mildred Peterson Foundation, National Science Foundation (SES 0094573) and the Kauffman Foundation.

1. A disclosure is a formal document that a faculty member files when it is believed that an invention with commercial potential has been made.

2. With regard to firm incentives to invest, Decheneux et al. (2003) provide evidence based on inventions licensed from MIT that the ability to appropriate returns via effective patent protection is an important determinant of whether or not firms will commercialize university inventions.

3. One half of the firms in our industry survey noted that they include delay of publication clauses in at least 90 percent of their university contracts. The average delay is nearly four months, with some firms requiring as much as a year's delay.

4. We started with 1983 in order to be well past the date of passage of the Bayh-Dole Act of 1980. Universities supplied us with data as far back as disclosure information could easily be retrieved. The 1996 end was for Purdue University. Purdue was the basis for our pilot study in this project, which was initiated in 1997; hence, we only collected data through 1996.

5. For many of the faculty we could not find the arrival and departure dates. However, for some of these we were able to confirm that they were on the faculty in a given year, even if we do not know arrival or departure dates, so they did not have to be dropped from the sample.

6. Since many faculty are listed in multiple departments, we use the average quality across all departments listing the faculty member.

7. According to the AUTM survey, most licenses executed at universities are in the life sciences.

8. For example, it is likely to be the case that publications and disclosure are simultaneously determined.

5

Commercialization and the Scientific Research Process

The Example of Plant Breeding

W. RONNIE COFFMAN, WILLIAM H. LESSER,
AND SUSAN R. MCCOUCH

INTRODUCTION

The nature of scientific research has changed rapidly in recent years. Applied biological research, such as plant breeding,[1] has changed more in the past fifteen years than at any time since the discovery of genetics. Change has been driven by discoveries in the field of biology and information technology and accelerated by laws protecting or patenting biological materials, including plant cultivars, genes, and numerous enabling technologies. Concerns exist that expansive patenting, in terms of number and breadth, may limit the future involvement of the public sector and jeopardize the utilization of our germplasm resources for the welfare of humanity (Barton 1997; Bragdon and Downes 1998; Kloppenburg 2004). Practicing plant breeders are very much aware of a major reduction in the free exchange of germplasm, although it has not been well documented. This chapter examines recent trends in the scientific research process using the example of plant breeding, discusses the forces that will shape the future, illustrates the interdependence between the public and private sectors, and concludes with projections about the future, particularly in the public sector.

RECENT TRENDS IN PLANT BREEDING

Frey (1996) conducted a survey of the change in science person years (SYs) devoted to plant breeding research and development (R&D) during the five-year period 1990–94. He found that the total number of SYs devoted to plant breeding R&D in the United States was 2,241 in 1994, distributed among private companies (1,499), State Agricultural Experiment Stations

(SAES) (529), the Agricultural Research Service (ARS) of the United States Department of Agriculture (USDA) (177), and Plant Materials Centers (PMCs) of the USDA (36). Over the five-year period the net loss of plant breeding SYs in SAESs was estimated to be 12.5, or 2.5 SYs per year. For the same period the growth in private industry SYs was 160, or about 32 SYs per year.

Many of the leading institutions in public-sector plant breeding have ceased to produce finished cultivars and are concentrating on methodology development, applying the discoveries of basic biology to crop improvement. Over the past few decades, the Agricultural Research Service of the United States Department of Agriculture (USDA/ARS) has chosen to concentrate on basic science, eliminating or curtailing nearly all of the cultivar development programs on minor crops. Today, USDA/ARS devotes only 12 percent of its plant breeding SYs to cultivar development (Frey 1997). State Agricultural Experiment Stations also reduced their collective human input into plant breeding by 2.5 SY per year during the period 1990–94.

Transgenic crops are rapidly gaining in importance, with the area increasing from 1.7 million hectares in 1996 to 58.7 million hectares in 2002 (James 2002). Transgenic crops are, of course, proprietary and are almost exclusively the product of the private sector in industrialized countries. It seems likely that this is just the beginning of sweeping changes in global food and fiber production based on genetic technologies. However, we must remember that the future of both public- and private-sector plant breeding ultimately depends on the perception of consumers worldwide. In "Playing God in the Garden," a *New York Times Magazine* article, Michael Pollan (1998) wrote:

> In a dazzling feat of positioning, the industry has succeeded in depicting these plants [transgenics] simultaneously as the linchpins of a biological revolution—part of a "new agricultural paradigm" that will make farming more sustainable, feed the world and improve health and nutrition—and, oddly enough, as the same old stuff, at least so far as those of us at the eating end of the food chain should be concerned.

Pollan termed this a "convenient version of reality" that has thus far been roundly rejected in Europe and ultimately could be rejected in the United States and other parts of the world. The more likely scenario would seem to be that it is just a question of time until these technologies are accepted worldwide, particularly once products with health or quality benefits for consumers become available over the coming decade. However, at this time the eventual acceptability of bioengineered foods for consumers is unknowable.

GENOMICS: FRAMING THE FUTURE

Biology, and the industries (agricultural and medical) that depend on it, is now involved in a revolution comparable to the one that took place in physics at the beginning of this century (Tanksley 1998). Large-scale DNA sequencing is revealing the genes required to encode most major life forms, including humans, microbes, plants, and animals. The scale of discovery is causing a major shift in the paradigm of biological research from a reductionist approach that focused on individual phenomena to a highly parallelized approach that integrates the molecular information for whole organisms across biological kingdoms and encompasses entire physiological and behavioral systems.

The major outcomes of genomics research over the next ten to thirty years will be: (1) the association of DNA sequence data with biological function and the determination of how nucleic acid and protein sequences have changed through evolutionary time to create the diversity of life forms that now inhabit this planet; and (2) understanding the flow and regulation of information encoded by the genome and the way genomic information is, in turn, regulated by information from the environment. The ensuing discoveries will revolutionize our understandings of the origins of life and the molecular processes that underlie life. They will also lead to many revolutionary discoveries in engineering, medicine, the environment, and agriculture.

As new discoveries in the genomics arena are applied, companies and industries are being restructured in a way that will change the world's economy (Enríquez 1998). Many of the world's largest companies have been forced to reinvent themselves as conventional demarcations blur. Initially, they formed what Enríquez termed the life sciences industry. He pointed out that the flow of genomics information is so massive that it threatened to overwhelm existing R&D budgets, labs, and knowledge bases.

Megamergers and dissolutions have been happening as companies attempt to lock in patents and licensing agreements and maximize profits. As investors began to recognize (a) the cost of bringing agricultural biotechnologies to market (in no small part due to regulatory costs), (b) the magnitude of public resistance to bioengineered crops (particularly in

Europe), and (c) the relatively limited returns to investment in agricultural research (people able to pay for food are already consuming more than they need), the vision of a life sciences industry has been abandoned. The highly profitable medical sector of such companies has been separated from the agricultural component, while the latter has reaffirmed its affiliation with the agricultural chemical industry as a source of cash flow for investment.

So far, the public sector has been mostly a spectator in this process. Interactions with the private sector have been largely in training scientists and as grant recipients for specifically defined products. As consolidation approaches its limit in the private sector, this is beginning to change. Large companies, as Lieberwitz notes in her chapter in this volume, are now seeking strategic alliances with public-sector institutions to gain access to new ideas. Comprehension of this prospect in the public sector is increasing, and the implications, particularly for plant breeding and most other areas of biological research, are extraordinary.

THE BAYH-DOLE ACT

The tremendous impact of commercialization on plant breeding research can be traced to the passage of the Bayh-Dole Act in 1980, a watershed for the licensing of innovative results that flowed from federally sponsored research projects in the United States. Bayh-Dole was the federal government's response to perception among governmental, industry, and academic leaders that few of the results of federally sponsored research projects were being commercialized and hence benefiting U.S. taxpayers. There was a broad sense at the time that the public was not receiving any significant benefit from its support of public research. Moreover, pressure existed both at the federal level and within many individual states to reduce public budgets, including the public support provided to higher education and its related research activities. Enabling research institutions to license the results of their researchers' work seemed a positive way to cushion the shock of reduced governmental research support in a way that respected the then-emerging "privatizing" atmosphere.

The key provisions of the Bayh-Dole Act[2] as described in Lieberwitz's chapter in this volume include:

uniform patent policy established for federally funded research;
university-industry collaboration encouraged;

universities and/or for-profit grantees/contractors[3] allowed to retain
   title to inventions developed through government funding; and
the government retained a nonexclusive license to practice the
   invention throughout the world (an option that could only
   be exercised if statutory protection was sought in foreign
   jurisdictions).

Bayh-Dole and subsequent guidelines have led to a dramatic increase in
university-industry intellectual property (IP) transfers and, overall, has been
considered highly successful in that regard.[4] Three key factors are credited
with that success:

certainty of title given to inventions;
leadership delegated to the inventors (individuals and/or
   institutions); and
uniform IP standard provided for all research conducted with
   government funds and a predictable patent and licensing
   procedure.

Because Bayh-Dole permitted research institutions to claim ownership
of federally funded research results, this law provided a mechanism for

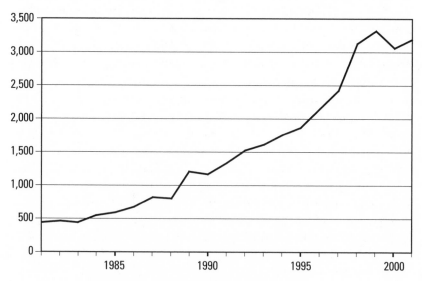

FIGURE 5.1 Patents awarded to U.S. universities, 1981–2001.

Source: National Science Board 2004, Table A5–54.

research institutions to commercialize those research results, and in this way, it was argued, the public would benefit. The terms of Bayh-Dole provided the incentive for many U.S. universities (and similar U.S. research institutions) to establish or expand their technology licensing activities. The increase in patents granted to U.S. colleges and universities following the enactment of Bayh-Dole is shown in Figure 5.1.

University Policies

The Bayh-Dole Act undoubtedly had a significant impact in the United States on the protection and transfer of university inventions. By extension, the impact on federally funded research on university IP policy and management/licensing of inventions strategies was considerable. At present, the U.S. government is one of a very few major research funders with an explicit IP policy. However, current discussions under several forums are shifting rapidly in the direction of establishing policies. That is, philanthropic foundations and other sponsor agencies are increasingly coordinating their policies to create a unified, level playing field. The objectives of these institutions mainly relate to their missions of ensuring access to research inventions by developing countries for social and humanitarian benefits.

This debate and ensuing policy shifts are expected to lead to new challenges for universities, especially ones with a long history and interest in collaboration with organizations in the developing world. These challenges are related to the difficulty of reconciling IP policies of different research sponsors and of harmonizing university IP policies with any possible new restrictions imposed.

A complementary development in the international debate relates to finding appropriate mechanisms for reducing the barriers that impede access specifically to agricultural biotechnology for subsistence and minor crops. Among others, what has been proposed is the creation of a clearinghouse to advise researchers, administrators, and technology managers about practical IP management strategies that will result in quicker decisions, lower transaction costs, and, ultimately, the biotechnological development and dissemination of plant varieties that address hunger (subsistence crops) or contribute to more vibrant state economies (minor commercial crops). In addition, endeavors are proceeding toward the creation of a mechanism such as a technology pool derived from many public-sector institutions to grant researchers broader access to complementary collections of agricultural biotechnologies, materials, and information for specific purposes.

Even with those processes in place, the obstacles to the transfer of modern biotechnologies to resource-poor farmers remain numerous. Included are:

covering regulatory costs;
partitioning markets with both commercial and small farmer
    applications;
developing transformation protocols for noncommercial crops like
    cassava;
provision of training;
establishing and monitoring oversight, such as may be needed for
    the delay of resistance development when using Bt crops; and
managing the tradeoff between costs and the maintenance of
    genetic diversity when multiple local varieties are being grown.

The question of allowing faculty to start a company (or take employment with a company) that licenses their own inventions from the university raises issues of conflict of interest. These issues are dealt with in the policy statements of many universities, and thus it is implicitly assumed that faculty have the right to such activity. But clear statements to this effect are not always found. Discussions about these matters in e-mail newsgroups address the question of allowing faculty "entrepreneurial leave of absence" to start their own companies, and it seems that faculty members often want to be involved in some way with the commercial development of their own inventions. The conflict of interest (and of commitment) issues need to be considered, but with a system in place to monitor this aspect, most universities allow, and often encourage, faculty to take such external roles. Indeed, the commercialization of technologies often requires the ongoing involvement of the inventors as those most knowledgeable about the potential of the product and the interested firms. In other cases, university technology transfer offices must take an equity position in a product to attract investors so that the offices' activities move beyond licensing into venture capitalism.

Existing university policies demonstrate apparent internal conflicts that lead to questions about their overall roles and effectiveness. Some examples include:

The university or its research foundation has a fiduciary
    responsibility that may best be exercised with exclusive licenses.
    Exclusive licenses can sometimes provide a lower public benefit

than nonexclusive licenses. The gene gun developed at Cornell is a good example (see Sanford et al. 1990; Klein et al. 1987). Conversely, exclusive licenses can be necessary to elicit the amount of additional investment required to bring a product to the commercialization stage.

Sometimes access to a technology is possible only by cross-licensing, involving other university inventions; but when doing that, a research foundation must make decisions about the use of technology that may have been discovered by another inventor(s). How should those conflicts be resolved under patent policies that promise researchers specific benefits following the licensing of their inventions?

Is it possible and appropriate to identify critical technologies (like the gene gun) and forgo exclusive licensing for greater public benefit? Does exclusive licensing actually provide for greater public benefit by providing adequate corporate incentive to expand the technology's use? These matters need to be studied in more detail and discussed openly so that policy outlines may be established.

Perhaps most significant to the research process, though, is the possible reduction in the exchange of information caused by property rights and financial benefit issues. Certainly there are adequate anecdotal examples of the chilling effects of property rights on information exchange and publication speed. But anecdotes are not a proper basis for policies. A few surveys have been conducted, particularly in the medicine/human genomics fields, that suggest some modest effects on information exchange and speed of publication that can be attributed to property rights practices. In many cases, property rights are claimed under contract law (MTAs) rather than patents so that modification of intellectual property rights (IPR) legislation, as has been proposed, will be insufficient to correct all the issues that have arisen.

More broadly from a policy point of view, the ongoing debate on policy shifts by other research sponsors, including the possibility for the creation of clearinghouses and patent pools (primarily based on U.S. university IP), is expected to have an impact on the way researchers will do business in the future. However, many key technologies (such as the Monsanto 35S promoter) are controlled by the private sector, which may not cooperate fully with the clearinghouses.

A faculty panel (Coffman et al. 2003) determined recently that Cornell University could better serve its internal and external responsibilities by placing a greater emphasis on the development and commercialization of university inventions. While such a change in emphasis would eventually involve many aspects of the university, a starting place was thought to be an increase in incentives for activities leading to and promoting commercialization. The following changes were recommended in university policy:

Recognize the issue of a patent on an invention as an academic contribution similar to the publication of a refereed journal article for promotion and tenure purposes;

Provide additional and, particularly, more rapid financial support (including for research) to inventors. The present system with a lag of five to eight years between invention and realization of any financial returns provides limited incentives for inventors to develop an invention further, particularly for younger professionals;

Modify the university Conflict of Interest policies to allow more joint activity as a university faculty or staff member and officer in a startup firm directed to commercializing the invention.

Participate in an effort by the Rockefeller Foundation and other leading research universities to establish an IPR clearinghouse and an IPR pool that will facilitate collective licensing of our technology for humanitarian use throughout the developing world.

The last recommendation recognizes that

IPR are here to stay and globalizing;

most key inventions will continue to occur in the public sector at research universities;

public funding should maximize public benefits, and food security is an important public benefit;

international agricultural research centers and national agricultural research systems throughout the world need help with access to IPR;

the private sector will not serve poor farmers;

private companies have IPR that they are willing to donate, and pooling IPR creates added value;

most university scientists would like to see their work benefit
   needy people; and
a portfolio of public IPR supplemented by case-by-case licensing
   can provide freedom to operate, and sharing that will benefit
   humanity.

However, university inventions are frequently at an early stage of development and require significant additional investment before usable products are available. This additional investment typically comes from the
private sector, which must be compensated for its expenditures and risk.
This is quite a different scenario from what is often involved in these discussions where university inventions are treated as being fully functional
right out of the laboratory.

Implications of Proprietary Technology

If present trends continue in the patenting of genes, it appears that two
or three companies will have a major influence on the global food system.
With the advent of proprietary technology, small seed companies will either
license technology from private industry suppliers or again look to the
public sector for advanced breeding material. But most public-sector programs have long since moved to more basic research and the development
of source material.

Because of the severely limited resources available in public-sector programs, the provision of advanced breeding material will be a necessary (if
small seed companies are to survive) but very limited opportunity. In fact,
with most important enabling technologies controlled by large-scale companies in the private sector, it is difficult to see how public-sector programs
can continue to be relevant in the production of advanced breeding material unless strong partnerships exist between public-sector programs and
those holding the enabling technologies in the private sector. Research
exemptions are generally available but leave the public-sector breeder in
the very difficult position of developing technology that she/he may not
be able to distribute.

Public/Private Interdependence

Education

The public and private sectors of plant breeding research genuinely need
each other. The public sector is in the best position to lead in the training

of plant breeders, now a lifelong endeavor. Hawk and Smith (1993) stress that applied corn breeding programs in the public sector will be necessary to develop the human resources essential to future breeding efforts. Industry can be a better partner and supporter of the public sector's education mission. Exploring new teaching techniques with industry involvement is highly desirable (McConnell 1997). Self-paced learning modules available over the Internet will be an important part of the continuing education of plant breeders. Plant breeding is changing dramatically, and we need a new vision and a new curriculum for training tomorrow's plant breeders. This training must consider the revolution in information technology as well as the revolution in molecular genetics. Overall, this situation may not be fundamentally different from other sectors, such as electronics, that have long been dominated by a few firms with large patent portfolios. The biggest difference is perhaps the lower profit margins in agriculture versus other sectors.

Minor Crops

The private sector will not be able to meet the needs in minor crops due to the limited opportunity for profit and will continue to look to the public sector to support such crops. Minor crops are defined as crops that (1) are cultivated on a limited acreage, (2) are produced as strains of major crops for niche markets, (3) provide relatively low gross revenues, (4) receive limited or no investment in research by either public or private sectors, and (5) feature plant breeding activities that are diffuse or nonexistent. A National Plan for Promoting Breeding Programs for Minor Crops in the U.S. has been developed (Frey 1997) by a coalition of private and public plant breeders, the Buckwheat Coalition. Elements of the plan include (1) establishing an organization to promote breeding of minor crops, the National Coalition for the Improvement and Use of Minor Crops (MCIC); (2) improving economic viability of minor crops; (3) promoting awareness of the significance of minor crops; and (4) securing sustained funding for minor crop breeding.

Some years ago, the pharmaceutical industry faced a similar issue for financing investments for rare diseases, the so-called orphan drugs. The response was the offering by the FDA of exclusive licenses for those products as well as grants targeted to their development.[5] While the exclusive license provisions in particular have been controversial, the approach does provide possible models for encouraging investment in the breeding of minor crops.

Genepool Enrichment

Genepool enrichment is a common need for all plant breeding programs but cannot be supported extensively by any individual program or company. The same public/private coalition concerned with minor crops has developed a National Plan for Genepool Enrichment of U.S. crops (Frey 1998). Elements of the plan include (1) establishing an organization to implement a crop genepool enrichment program, the Crop Genepool Enrichment Coalition (CGEC); (2) providing for the timely and orderly enrichment of genepools of U.S. crops; (3) promoting awareness of the significance of crop genepool enrichment to the future viability of U.S. agricultural production; and (4) securing sustained funding for crop genepool enrichment.

At the international level, the 2001 FAO International Treaty on Plant Genetic Resources for Food and Agriculture has established a standardized agreement for accessing genetic resources for designated crops in public genebanks. The initial and essentially derived variety provisions of the 1991 UPOV act are intended to provide incentives for background breeding. The United States adopted those provisions in 1994,[6] although they have yet to be implemented here or elsewhere.

THE FUTURE OF PLANT BREEDING RESEARCH

Beginning early in this century and peaking with the Green Revolution, the stream of public research has been of enormous benefit to humanity, particularly the citizens of the United States, who have spent an ever-diminishing portion of their income on food. The majority of this benefit came from plant breeding (Ruttan 1982) and was based on an implicit contract that genetic resources were public and available to anyone. Lately the rules have changed (Kloppenburg 2004). More and more of the best science is patented. Some scientists believe that any DNA of value will eventually become somebody's property (Herdt 1999). Others suggest that only discrete sequences of DNA that represent a valuable asset will be patented. In either case, the beneficiaries will be the stockholders and employees of the private sector and consumers who will enjoy lower food prices. Producers are expected to move increasingly to contract production for products developed specifically for specialized markets. Contract production can have mixed effects, from reducing management requirements to shifting some risks onto the contractor. Producers with the poorest soil/water resources can be expected to suffer the most if denied access to cost-reducing technologies, while crop prices fall as producers elsewhere, in more favorable circumstances, adopt those same technologies. That is essentially what occurred with past Green Revolution technologies. However, in some cases, new technology (drought tolerance, for instance) may allow enhanced use of those resources.

Plant breeding will become integrated fully with agricultural research specifically and with biological, engineering, and medical research generally. Future plant breeders will rely heavily on computers to an extent that can hardly be imagined today. A plant is a biological entity driven by a piece of code interacting with the biological, physical, and chemical environment. That code will be deciphered at a rapidly increasing rate, along with the code for other organisms in the environment, particularly pest organisms. All useful code eventually may be proprietary. The characterization of production environments will be refined continuously. It is unlikely, in the fullness of time, that the public sector will have a role that is independent of the private sector. Indeed, even today, public-sector agricultural research would be much diminished but for the access and contracts provided by private firms.

Change is coming quickly in some areas. The major firms in the private sector are interested in partnerships with the public sector that offer a "first look" at new ideas. This became evident when *U.S. News and World Report* (Petit 1998) reported that the University of California, Berkeley, had completed "an unprecedented deal to sell access to an entire department. For $25 million up-front for new campus laboratories and $25 million in research funds over the next five years, the Swiss-based biotech giant, Novartis,[7] gets to observe the work of 32 faculty members and nearly 200 graduate students and postdoctoral fellows in the Department of Plant and Microbial Biology. Novartis also gets first crack at negotiating the rights to take the department's discoveries to market."

Can we expect more alliances like this in the future as well as increasing alliances among competitors in the private sector? Low current profits and investor disillusionment with agricultural biotechnology means large upfront payments are unlikely at present. Numerous lawsuits now pending on biotech patents will likely be settled so that such patents are shared among "Haves." "Have-Nots" will be left to license technology at increasing royalty rates. The public sector may tend to partner with the "Haves" because public-sector breeders generally are starved for resources and need such partnerships to maintain their education programs. The true implications of domination by large firms depend in no small part on the rate of the advancement of the science. If it continues at a high rate, then rapid obsolesce means protection is relatively less important than when advances are more paced.

Unification of the public sector may become a significant factor. A group of research university presidents (Demaine and Fellmeth 2003) is attempting to unite public-sector institutions to conserve the right to utilize their collective intellectual property for humanitarian purposes. They have expressed their intent to form the Public-Sector Intellectual Property Resource for Agriculture (PIPRA) for the purpose of (1) reviewing public-sector licensing practices; (2) creating a collective, public, IP asset database; (3) pooling specific IP to form shared technology packages; and (4) inviting broad participation by other public-sector institutions.

The future of plant breeding depends on the policies that we evolve for the management of intellectual property related to crop improvement. R.G. Sears (1998) pointed out that the free exchange and utilization of germplasm has been the foundation for all plant improvement efforts since crops were first cultivated more than 10,000 years ago. He draws the analogy to a book (which may be copyrighted) and words (which may not be copyrighted). If words were copyrighted, only the few who owned them could communicate and our society would be harmed. Genes are analogous to words in that they allow the creation of new plant cultivars just as words allow the creation of a new book. Everyone in society should have the right to use genes. Cultivars (novel genotypes or combinations of genes), not genes, should be eligible for patenting. It is now clear that the current system of patenting genes restricts the playing field so that only two or three companies will have a major influence on the global food system. In a broader context, the debate over the patentability of "discoveries" such as naturally occurring genes is an age-old one. Key interpretations are now being made by the U.S. Patent and Trademark Office (Cowley and Makowski 2003).

In the end, a decision probably will be made based on what makes economic sense. In a 2003 presentation reported by *BBC News*, Alan Greenspan mulled the issues of patenting:

Mr. Greenspan made no judgments over whether the rules designed to cope with real world, concrete property were too tight or too loose to make sure protection for ideas contributed to economic growth. Nor did he comment on the Fed's economic policies or the state of the U.S. economy. But he made it clear that hard decisions and deep thinking were needed to make sure that the balance was right. "Ownership of physical property is capable of being defended by police, the militia, or private mercenaries," he said. "Ownership of ideas is far less easily protected." One ancient example, he commented, was that of calculus, the mathematical tool discovered by Gottfried Leibniz and Sir Isaac Newton in the 17th Century, noting that their discovery was made freely available, triggering massive intellectual advances elsewhere. "Should we have protected their claim in the same way that we do for owners of land?" he asked. "Or should the law make their insights more freely available to those who would build on them, with the aim of maximizing the wealth of the society as a whole? Are all property rights inalienable, or must they conform to a reality that conditions them?"

In his recently revised book *First the Seed* (2004), Jack Kloppenburg emphasizes what he sees as the four principal reasons for maintaining a robust public presence in plant improvement:

The public sector can produce public goods that are socially valuable but do not attract private investment because they are not profitable.
By developing and releasing finished cultivars, the public sector can provide a countervailing force to the market power of large companies as well as maintain standards of quality.
The public sector can explore options and alternatives and innovations that are not pursued by private industry.
The public sector can supply independent and reliable information resulting in an informed populace capable of making informed decisions in a complex and uncertain world.

These are laudable points that most everyone in the public, as well as the private, sector would likely support. However, resources to support the public sector are unlikely to reappear, especially in quantities sufficient to support state-of-the-art crop improvement programs and the regulatory packages that may be necessary to take some of the products to market. Opportunities exist for public-private partnerships whereby the private sector can share technology without sacrificing market share and the public sector can use it to benefit the resource-poor. The Agricultural Biotechnology Support Project II (http://www.absp2.cornell.edu/) at Cornell University is one example of a successful effort to form such partnerships.

NOTES

1. Plant breeding is a relatively unrecognized scientific specialty that is the basis of human civilization. Fewer than 2500 plant breeders practice in the United States—fewer than 7,500 (estimated) worldwide—sustaining a relatively small number of crops that capture solar energy and sustain human existence. Rice, for instance, is the primary staple of half the world's population. The rice genome was first mapped at Cornell University (McCouch et al. 1988) Cornell supports the only Department of Plant Breeding in the United States, numbering twelve faculty members.

2. P.L. 96–517 of 12 December 1980 and subsequent modifications (P.L. 98–620 of 8 November 1984).

3. The for-profit grantee clause was part of an amendment in the form of a Presidential Memorandum on *Government Patent Policy* of 18 February 1983.

4. Mowery et al. (2001, 2) suggest that Bayh-Dole did not cause the dramatic increase in university patent activity but rather that the "principal effect of Bayh-Dole was to accelerate and magnify trends that already were occurring in academe."

5. See http://www.fda.gov/orphan/.

6. Plant Variety Protection Act of 1994, Sec. 111(c).

7. Subsequent to this agreement, the agricultural component was spun off from Novartis and merged with a similar spin-off to form Syngenta.

Foreign Students
and Scholars

6

The Importance of Foreign
Ph.D. Students to U.S. Science

GRANT C. BLACK AND PAULA E. STEPHAN

INTRODUCTION

During the 1980s and 1990s, science and engineering (S&E) Ph.D. programs in the United States became increasingly populated with and dependent on foreign students. Fueled by the large increase in noncitizen students, Ph.D. programs grew by 61.7 percent during the period 1981–99. If U.S. Ph.D. programs had grown at the citizen doctorates rate instead, they would have increased by only 26.3 percent. Postdoctoral programs have also become increasingly populated by noncitizens. For example, 37 percent of individuals on postdoctoral appointments in the United States were on temporary visas in 1982; by 2002 the percent stood at 59 (National Academy of Sciences 2005b).

The objective of this study is three-fold: (1) to document these trends in Ph.D. production, examining them by field and by country of origin; (2) to explore how the increase in the foreign-born Ph.D. population relates to the selectivity of Ph.D. programs; and (3) to examine factors affecting the rate at which foreign-born Ph.D. recipients on temporary visas stay in the United States. The third point is of particular interest since it impacts the composition of the U.S. workforce and can affect the flow of knowledge from the United States to other countries.

Data for the study come from the Survey of Earned Doctorates (SED), administered by Science Resources Statistics (SRS) of the National Science Foundation (NSF). The survey is a census of all doctoral recipients in the United States and has a response rate of approximately 92 percent.[1]

We use individual level data made available to us through a license with Science Resources Statistics, National Science Foundation. We restrict our study to individuals in sixteen fields of science and engineering, purposely excluding those trained in the humanities as well as the social sciences, economics, business, and psychology. Unless noted, our focus is on those who hold a temporary visa at the time of receipt of the degree.

Trends

During the period 1981–99, temporary residents accounted for more than 50 percent of the growth in Ph.D. production in the United States (Stephan et al. 2002). Permanent residents provided for another 10 percent. Growth of the foreign born was especially strong during the first twelve years of the period. The number of foreign born declined somewhat during the early 1990s but increased toward the end of the 1990s.

Figure 6.1 documents the dramatic increase in the number of Ph.D. recipients holding temporary visas during the period 1981–92, followed by a decline during the next seven years. In 1981 fewer than 2,500 Ph.D. recipients in S&E held temporary visas (20 percent of all those receiving Ph.D.s in S&E); by 1992 the number stood at close to 7,000 (38.4 percent of all doctoral degrees awarded in S&E that year). Seven years later, in 1999, the number had decreased by approximately 1,000, with temporary visa recipients receiving approximately 32 percent of all Ph.D.s awarded in S&E that year. Part of the decline in the early to mid-1990s reflects the passage of the Chinese Student Protection Act that permitted Chinese nationals temporarily residing in the United States to switch to permanent resident status. The decline is also related to a statistical artifact. Beginning in 1997,

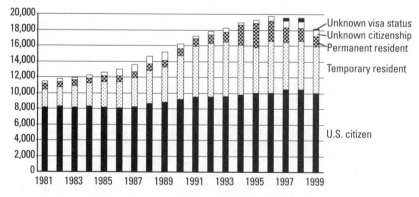

FIGURE 6.1 Citizenship status of S&E doctorates by year of degree, 1981–99.

when the SED survey procedures were changed, a considerable increase in the number of doctorate recipients with "unknown" citizenship status occurred (see Figure 6.1).

The growth in temporary residents has been especially dramatic in the fields of the biological and agricultural sciences and math and computer sciences. In the biological and agricultural sciences, the percent of temporary residents receiving Ph.D.s more than doubled during the period 1981–92, going from approximately 13 percent to almost 28 percent. It then fell slightly, to approximately 26 percent by 1999. In math and computer sciences, the percent increased from 23.5 percent in 1981 to 46 percent in 1991; it stood at 39 percent in 1999. The change in composition has been less dramatic in engineering, but the proportion of doctorate recipients who are temporary residents in this field is substantial, hitting a high of 50.5 percent in 1991 and closing the decade at 39.6 percent.

Country of Origin

The country of citizenship of doctorate recipients with temporary visas for the decade of the 1990s is indicated on the map in Figure 6.2.[2] Particularly striking is the large concentration of recipients from Asia, with 60 percent coming from four countries: the People's Republic of China (21.0

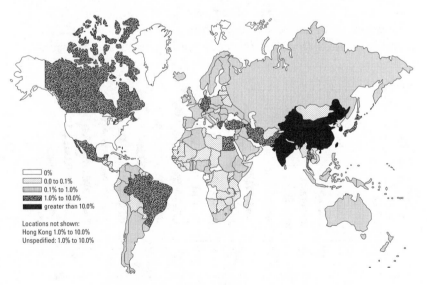

FIGURE 6.2 Proportion of temporary residents receiving S&E doctorates in the United States in the 1990s by country of citizenship.

percent); Taiwan (13.7 percent); India (12.2 percent); and South Korea (11.1 percent).[3] Equally striking is the fact that the next six most frequent countries are geographically dispersed (Canada, Brazil, Turkey, Greece, Germany, and Mexico). Moreover, recipients from these six countries collectively make up less than 11 percent of doctoral recipients with temporary visas. Indeed, the distribution is so highly skewed that no countries fall in the range of 3 percent to 11 percent.

Selectivity of Ph.D. Programs

Figure 6.3 shows the distribution of temporary residents earning S&E Ph.D.s in the United States during the period 1981–99 by top-ten- and non-top-ten-ranked Ph.D. programs. Rankings, in almost all instances, are taken from the 1993 National Research Council study.[4] After a gradual increase through the early 1990s, followed by a slight dip, the number of temporary residents from top programs remained fairly constant, at around 1,000, during the remainder of the period. Because of the growth in foreign students during this period, the proportion of temporary residents graduating from top-ranked programs fell considerably. In 1981, about a quarter (23.4 percent) of temporary resident Ph.D. recipients graduated from a top-ten program; by 1999 this had dropped to 16.1 percent. It is not clear why this decrease occurred. Possible explanations include a change in the mix of program interest on the part of temporary residents; a change in quality of the applicant pool, or the adoption of an implicit quota regarding the number of temporary residents in elite programs during the 1990s.[5]

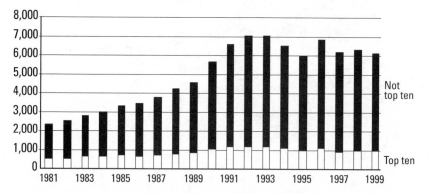

FIGURE 6.3 Number of temporary resident Ph.D.s by selectivity of institution, 1981–99.

Source: National Research Council 1995.

Although only about one in six temporary residents in S&E graduate from a top-ten program, the proportion varies considerably by field. Table 6.1 shows the percentage of doctorate recipients during the period 1981–99 who received their degrees from highly selective programs by field and citizenship status. We see that more than 50 percent of temporary residents in oceanography (51.6 percent) and aerospace engineering (50.9 percent) graduated from top-ten programs in their fields, while only 10.3 percent of temporary residents in chemistry and 6.8 percent of temporary residents in biology earned degrees from top-ten programs.

It does not follow that temporary residents are less likely to graduate from highly selective programs than their citizen and permanent resident counterparts. Indeed, there is little difference between temporary residents and U.S. citizens in the overall proportion of graduates from top programs. Over the period analyzed, 18 percent of temporary residents graduated from highly selective programs compared to 20 percent of U.S. citizens. However, differences emerge across fields between temporary residents and citizens. In the fields of chemistry, biological sciences, and physics, citizens are almost twice as likely to receive their degree from a top program than

TABLE 6.1 Doctoral education at top-ten programs, 1981–99.

Field	Percent of all doctorate recipients	Percent of temporary residents	Percent of U.S.citizens
Aerospace engineering	47.6	50.9	44.4
Agriculture	29.6	29.2	30.1
Astronomy	37.8	40.4	37.1
Biological sciences	10.7	6.8	11.4
Chemical engineering	27.7	19.3	36.6
Chemistry	17.4	10.3	20.1
Civil engineering	27.4	27.2	27.9
Computer sciences	22.6	19.3	24.8
Earth sciences	20.8	18.3	20.7
Electrical engineering	28.7	23.1	32.3
Mathematics	20.7	19.6	20.1
Mechanical engineering	27.5	23.7	32.2
Medicine	16.4	14.3	17.3
Oceanography	53.7	51.6	52.7
Other engineering	17.3	13.4	19.2
Physics	22.9	15.9	27.0
All Science & Engineering	19.9	18.1	20.5

are temporary residents. In other fields (aerospace engineering and astronomy) temporary residents are more likely than citizens to attend a top-rated program.

Stay Patterns

The U.S. scientific doctoral workforce has become increasingly foreign born (Levin and Stephan 1999; Stephan and Levin 2001; Stephan and Levin, chap. 8 herein). While some doctoral scientists immigrate to the United States after receiving their Ph.D.s abroad, many come for doctoral study and stay. Moreover, this pattern has increased during the past twenty years (Finn 2000). Increased stay rates, coupled with increased degrees awarded to individuals on temporary visas, have brought about an extraordinary increase in the number of foreign nationals who receive their degrees and then work in the United States. For example, the percentage increase in doctorates awarded to temporary residents who were in the United States three to four years after their degree was 231 percent in the life sciences, 131 percent in the physical sciences, and 93 percent in engineering between the 1987–88 period and the 1992–93 period (Finn 2000, 4). During the comparable period, U.S. citizen doctorate recipients increased by only 0.9 percent in the physical sciences, 11 percent in the life sciences, and 29.9 percent in engineering.

Figure 6.4 shows the percent of doctoral recipients on temporary visas who report that they plan to stay in the United States at the time they

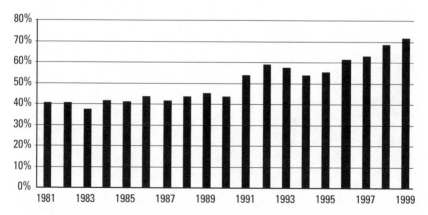

FIGURE 6.4 Proportion of S&E temporary residents with plans to stay in the United States, 1981–99.

receive their Ph.D.s. We see that the increase in stay rates was particularly noticeable, though uneven, during the period 1991–99.

Stay patterns vary considerably by country of origin, as seen in Figure 6.5. Among the top-ten sending countries, for example, we find significant differences in stay patterns. For the big four, those from China and India report the highest plans to stay; those from Taiwan and South Korea are significantly lower. Among the other top-ten countries, stay patterns are lowest for those from Brazil and Mexico.

These definitions of stay patterns are based on the respondent's answer to a question regarding location plans subsequent to graduation. Specifically, the question on the SED asks doctoral recipients to "name the organization and geographic location where you will work or study" for those indicating definite plans or to identify "in what state or country you intend to live after graduation" for all others. Although the question undoubtedly invites an optimistic response, Finn's (2000) research indicates that 53 percent of all students on temporary visas are in the United States three to four years after receipt of their degree. While Finn does not test to see how this relates to "stay plans," the stay plans reported in Figure 6.4 are reasonably consistent with Finn's findings, especially when social scientists, who have the lowest stay rates, are eliminated from Finn's findings. Once this group is eliminated, the stay patterns that Finn reports vary between a low of 50 percent and a high of 62 percent, depending on field and cohort.

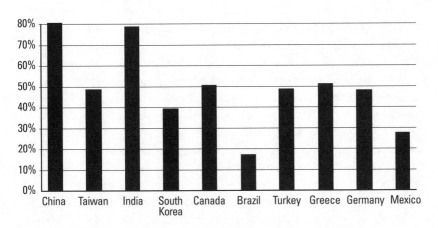

FIGURE 6.5 Stay rate for temporary residents from top-ten countries, 1981–99.
Source: NSF, Survey of Earned Doctorates.

In an effort to understand underlying factors affecting stay plans, we estimate a logit model for temporary residents receiving Ph.D.s in sixteen fields of S&E who indicate their stay plans in the survey for the period 1981–99. (The reader can find a brief discussion of the logit technique in the statistical section of the introduction to this volume.) Missing from the analysis are 20,074 of the 94,474 temporary residents who received a doctorate in S&E during this period: 14,209 individuals who do not reply to the postdoctoral location question, 469 individuals with an S&E degree not in one of the sixteen fields, and 5,396 individuals for whom there are missing observations on the independent variables of interest.[6]

Variables are defined in Table 6.2, and means and standard deviations are presented. The logit results are presented in Table 6.3. The omitted field is biology, and the omitted countries are non-top-tens.[7] In addition to showing the logit coefficients and levels of significance, we report the marginal effect, evaluated at the means, of a change in the independent variable. In the case of a dummy variable, these marginal effects show by how much the probability will change with a change in status; in the case of a continuous variable, they show how much the probability will change with a one-unit change in the value of the variable.

We find that demographics play a decisive role in determining stay patterns: age and marital status matter. Consistent with human capital theory, stay patterns decrease (and at a decreasing rate) with age. Married individuals are less likely to stay, possibly reflecting the challenge of finding two positions in the United States as well as the spouse's pull to return to the country of origin.

Ties to the United States also dramatically affect stay plans. Particularly noticeable is the strong positive impact of having received one's bachelor's degree in the U.S., which increases the probability of staying by approximately 0.10. Work experience in the U.S. also plays an important role. Those who were working full time the year prior to receiving their Ph.D.s are significantly more likely to plan to stay than those not employed, as are those who were working part time. Being supported by a fellowship also significantly increases the likelihood that one plans to stay compared to those not employed. A plausible explanation is that individuals on fellowships build stronger networks with U.S. researchers than those who are not on fellowships. There is also the related factor that individuals on fellowships are selected for their ability and recipients realize that this signals U.S. employers concerning their quality. Those who report that they were involved in the catch-all "other" predoctoral status category are significantly

TABLE 6.2 Descriptive statistics.

Variable	Definition	Mean (Standard deviation)
Stay	Dummy variable indicating whether or not an individual has intentions to stay in the U.S. regardless of the definiteness of those plans	0.64 (0.48)
Defstay	Dummy variable indicating whether or not an individual has definite plans to stay in the US, based on plans to return to or continue in predoctoral employment, or negotiations or a contract with a specific employer	0.44 (0.50)
Age	Age of the individual at time of Ph.D.	33.0 (4.4)
Age Squared	Age of the individual squared	1109.5 (312.4)
Female	Dummy variable indicating whether or not an individual is female	0.16 (0.36)
Married	Dummy variable indicating whether or not an individual was married at time of Ph.D.	0.62 (0.49)
Preftemp	Dummy variable indicating whether or not an individual was full-time employed one year prior to Ph.D.	0.27 (0.44)
Prefellow	Dummy variable indicating whether or not an individual was supported by fellowships or assistantships one year prior to Ph.D.	0.54 (0.50)
Preptemp	Dummy variable indicating whether or not an individual was part-time employed one year prior to Ph.D.	0.07 (0.25)
Prenotemp	Dummy variable indicating whether or not an individual was not employed one year prior to Ph.D.	0.09 (0.29)
Preother	Dummy variable indicating whether or not an individual held any other predoctoral status one year prior to Ph.D. (including unknown status)	0.03 (0.18)
U.S. Bachelors	Dummy variable indicating whether or not an individual received a bachelor's degree from an institution in the United States.	0.07 (0.25)
China	Dummy variable indicating whether or not an individual was a citizen of China at time of Ph.D.	0.16 (0.37)
Taiwan	Dummy variable indicating whether or not an individual was a citizen of Taiwan at time of Ph.D.	0.14 (0.35)
India	Dummy variable indicating whether or not an individual was a citizen of India at time of Ph.D.	0.12 (0.33)
South Korea	Dummy variable indicating whether or not an individual was a citizen of South Korea at time of Ph.D.	0.10 (0.30)
Canada	Dummy variable indicating whether or not an individual was a citizen of Canada at time of Ph.D.	0.03 (0.17)

Table 6.2—*Continued*

Variable	Definition	Mean (Standard deviation)
Brazil	Dummy variable indicating whether or not an individual was a citizen of Brazil at time of Ph.D.	0.02 (0.15)
Turkey	Dummy variable indicating whether or not an individual was a citizen of Turkey at time of Ph.D.	0.02 (0.13)
Greece	Dummy variable indicating whether or not an individual was a citizen of Greece at time of Ph.D.	0.02 (0.13)
Germany	Dummy variable indicating whether or not an individual was a citizen of Germany at time of Ph.D.	0.01 (0.12)
Mexico	Dummy variable indicating whether or not an individual was a citizen of Mexico at time of Ph.D.	0.02 (0.13)
Aere	Dummy variable indicating whether or not an individual's Ph.D. field was aerospace engineering	0.01 (0.12)
Chee	Dummy variable indicating whether or not an individual's Ph.D. field was chemical engineering	0.05 (0.21)
Cive	Dummy variable indicating whether or not an individual's Ph.D. field was civil engineering	0.05 (0.22)
Elee	Dummy variable indicating whether or not an individual's Ph.D. field was electrical engineering	0.10 (0.30)
Mece	Dummy variable indicating whether or not an individual's Ph.D. field was mechanical engineering	0.06 (0.24)
Oeng	Dummy variable indicating whether or not an individual's Ph.D. field was another engineering field	0.12 (0.32)
Astr	Dummy variable indicating whether or not an individual's Ph.D. field was astronomy	0.01 (0.07)
Chem	Dummy variable indicating whether or not an individual's Ph.D. field was chemistry	0.10 (0.30)
Phys	Dummy variable indicating whether or not an individual's Ph.D. field was physics	0.08 (0.27)
Eart	Dummy variable indicating whether or not an individual's Ph.D. field was earth sciences	0.02 (0.15)
Ocea	Dummy variable indicating whether or not an individual's Ph.D. field was oceanography	0.004 (0.07)
Math	Dummy variable indicating whether or not an individual's Ph.D. field was mathematics	0.07 (0.26)
Comp	Dummy variable indicating whether or not an individual's Ph.D. field was computer sciences	0.05 (0.21)
Agri	Dummy variable indicating whether or not an individual's Ph.D. field was agricultural sciences	0.09 (0.28)

TABLE 6.2—*Continued*

Variable	Definition	Mean (Standard deviation)
Biol	Dummy variable indicating whether or not an individual's Ph.D. field was biological sciences	0.15 (0.36)
Medi	Dummy variable indicating whether or not an individual's Ph.D. field was medicine	0.04 (0.19)
Topaere	Dummy variable indicating whether or not an individual's Ph.D. institution was ranked in the top ten for aerospace engineering	0.007 (0.08)
Topchee	Dummy variable indicating whether or not an individual's Ph.D. institution was ranked in the top ten for chemical engineering	0.01 (0.01)
Topcive	Dummy variable indicating whether or not an individual's Ph.D. institution was ranked in the top ten for civil engineering	0.01 (0.12)
Toplee	Dummy variable indicating whether or not an individual's Ph.D. institution was ranked in the top ten for electrical engineering	0.02 (0.15)
Topmece	Dummy variable indicating whether or not an individual's Ph.D. institution was ranked in the top ten for mechanical engineering	0.01 (0.12)
Topoeng	Dummy variable indicating whether or not an individual's Ph.D. institution was ranked in the top ten for other engineering fields combined	0.01 (0.12)
Topastr	Dummy variable indicating whether or not an individual's Ph.D. institution was ranked in the top ten for astronomy	0.002 (0.05)
Topchem	Dummy variable indicating whether or not an individual's Ph.D. institution was ranked in the top ten for chemistry	0.01 (0.10)
Topphys	Dummy variable indicating whether or not an individual's Ph.D. institution was ranked in the top ten for physics	0.01 (0.12)
Topeart	Dummy variable indicating whether or not an individual's Ph.D. institution was ranked in the top ten for earth sciences	0.004 (0.07)
Topocea	Dummy variable indicating whether or not an individual's Ph.D. institution was ranked in the top ten for oceanography	0.002 (0.05)
Topmath	Dummy variable indicating whether or not an individual's Ph.D. institution was ranked in the top ten for mathematics	0.02 (0.12)
Topcomp	Dummy variable indicating whether or not an individual's Ph.D. institution was ranked in the top ten for computer sciences	0.009 (0.10)
Topagri	Dummy variable indicating whether or not an individual's Ph.D. institution was ranked in the top ten for agricultural sciences	0.02 (0.16)
Topbiol	Dummy variable indicating whether or not an individual's Ph.D. institution was ranked in the top ten for biological sciences	0.01 (0.10)
Topmedi	Dummy variable indicating whether or not an individual's Ph.D. institution was ranked in the top ten for medicine	0.005 (0.07)
Ph.D. Year	Year the individual received his Ph.D.	1992 (4.87)

TABLE 6.3 Logit estimation of the stay rate.

Variable	STAY		DEFSTAY	
	Estimated coefficient	Marginal effect	Estimated coefficient	Marginal effect
Intercept	−25.81	−5.29	37.94	9.29
Age	−0.14‡	−0.029	−0.15‡	−0.036
Age Squared	0.00091‡	na	0.0010‡	na
Female	0.041	0.0085	−0.097‡	−0.024
Married	−0.099‡	−0.021	0.0086	0.0021
Preftemp	0.43‡	0.10	1.21‡	0.26
Prefellow	0.77‡	0.17	1.22‡	0.26
Preptemp	0.70‡	0.16	0.86‡	0.18
Preother	0.25‡	0.059	0.22‡	0.039
U.S. Bachelors	0.52‡	0.098	0.32‡	0.080
China	2.61‡	0.39	0.94‡	0.23
Taiwan	0.34‡	0.083	0.19‡	0.044
India	1.84‡	0.33	1.04‡	0.25
South Korea	−0.041	−0.010	−0.0012	−0.00030
Canada	−0.089	−0.022	0.17‡	0.040
Brazil	−1.22‡	−0.28	−1.07‡	−0.20
Turkey	0.026	0.0064	0.086	0.020
Greece	0.025	0.0062	0.097	0.023
Germany	−0.28‡	−0.070	−0.0044	−0.0010
Mexico	−0.62‡	−0.15	−0.49‡	−0.10
Aere	−0.61‡	−0.12	−1.05‡	−0.25
Chee	−0.46‡	−0.091	−0.66‡	−0.16
Cive	−0.69‡	−0.14	−0.97‡	−0.23
Elee	−0.26‡	−0.048	−0.51‡	−0.13
Mece	−0.51‡	−0.10	−0.82‡	−0.20
Oeng	−0.51‡	−0.10	−0.71‡	−0.18
Astr	−0.26*	−0.049	−0.33†	−0.083
Chem	0.13‡	0.022	−0.11‡	−0.027
Phys	−0.21‡	−0.038	−0.35‡	−0.086
Eart	−0.62‡	−0.13	−0.59‡	−0.15
Ocea	−0.94‡	−0.20	−0.93‡	−0.22
Math	−0.53‡	−0.11	−0.54‡	−0.13
Comp	−0.39‡	−0.075	−0.43‡	−0.11
Agri	−1.20‡	−0.27	−1.24‡	−0.29
Medi	−0.62‡	−0.13	−0.54‡	−0.13
Topaere	0.22	0.048	0.38‡	0.085
Topchee	0.12	0.025	0.48‡	0.12
Topcive	−0.067	−0.016	0.16*	0.035
Topelee	0.31‡	0.057	0.43‡	0.11
Topmece	0.27‡	0.057	0.32‡	0.075
Topoeng	−0.031	−0.0069	0.029	0.0068
Topastr	−0.32	−0.068	0.062	0.015

TABLE 6.3—*Continued*

| | STAY | | DEFSTAY | |
Variable	Estimated coefficient	Marginal effect	Estimated coefficient	Marginal effect
Topchem	0.084	0.013	0.44‡	0.11
Topphys	−0.035	−0.0068	0.29‡	0.072
Topeart	0.10	0.024	0.096	0.023
Topocea	0.64‡	0.15	0.74‡	0.18
Topmath	0.062	0.014	0.34‡	0.084
Topcomp	0.39‡	0.076	0.69‡	0.17
Topagri	−0.089	−0.022	−0.061	−0.012
Topbiol	0.23†	0.038	0.23‡	0.057
Topmedi	−0.40‡	−0.095	−0.45‡	−0.10
Ph.D. Year	0.015‡	0.0031	−0.018‡	−0.0043

Note: n = 74,400
* Significant at the 10% level
† Significant at the 5% level
‡ Significant at the 1% level

more likely to plan to stay than those not employed, but the marginal effect is much smaller than for those who were employed or on a fellowship.

Stay plans, as indicated earlier, also depend on nationality. The probability of a Chinese student staying is 0.39 higher than is the probability of a student coming from a non-top-ten source country. The probability of those from India staying is 0.33 higher; for those from Taiwan it is 0.08 higher. Stay patterns for those from South Korea are not significantly different from the benchmark. Stay patterns from the two major sending countries south of the border, Brazil and Mexico, are considerably lower.

Stay plans are directly related to field of training. This is not surprising, especially since the ability to stay in the United States if one holds a temporary visa at the time of receipt of the degree depends in part on one's ability to receive a work permit or training permit. In certain fields work permits are more easily obtained than in others. This was especially the case during the IT boom of the mid-to-late 1990s. Also, work visas are not required for individuals planning to take a postdoctoral position, since such positions qualify as training.[8] Broadly speaking, these postdoctoral positions are most likely to be filled by those trained in the field of biology, the benchmark field in the estimated equation. The raw data reflect this fact. The highest overall stay plans of any discipline occur in biology, where 73 percent indicate that they plan to reside in the United States after finishing their degrees.

Computer science and electrical engineering were two fields in high demand during the 1990s and two fields in which H1B visas were often issued. It is therefore not surprising to find that stay plans in these two fields are only slightly lower than in the postdoctoral-oriented biological sciences, where stay patterns are particularly high. We find those trained in chemistry to be even more likely to plan to stay than those in the biological sciences. This undoubtedly reflects the dual opportunities available to chemistry Ph.D.s of either going to industry or of taking a postdoc position.

By way of contrast, those receiving Ph.D.s in the earth sciences and in oceanography are considerably less likely to stay, relative to the benchmark, as are those trained in agriculture and medicine. The earth science and oceanography results may well reflect the fact that the United States does not enjoy the dominant position in these fields worldwide. Indeed, these are two fields that some U.S.-born scientists consistently choose to receive their doctoral training in outside the U.S. The strong agricultural result may be consistent with the fact that source countries invest in the training of scientists in agriculture with the expectation that they will return home. Finn (2000) also finds low actual stay rates among those trained in agriculture.

The field results are affected in many instances by the quality of the Ph.D. program where the training was received. Using the same definitions developed above, we find that those trained in a top-ten electrical engineering program as well as a top-ten computer science program are more likely to stay than are those trained in non-top-ten programs in these fields. This is consistent with U.S. demand being higher for individuals from strong programs as well as the willingness of potential employers to seek visas for exceptionally well-trained temporary residents. We also see that those trained at top mechanical engineering programs and top biology programs are more likely to plan to stay. Interestingly enough, we also find that those from top oceanography programs are more likely to stay.

Overall, the quality results are consistent with the findings of Stephan and Levin (2001; and chap. 8 herein), which suggest that the foreign-born who work in the United States represent a highly select group. While Stephan and Levin focus on the selectivity process that draws high-achieving students to the United States, these results suggest that it is not only selection in terms of who comes but also selection in terms of who stays.[9]

The SED not only ascertains the plans of individuals at the time of graduation; it also ascertains whether individuals have "definite plans," meaning, whether the individual has plans to return to or continue in predoctoral employment or is negotiating or already has a contract with a specific employer. Using this more restrictive definition, we find that 44 percent of temporary residents who received their Ph.D. during the period 1981–99 have "definite plans to stay," compared to 64 percent with "plans to stay." Column five of Table 6.3 reports the marginal effects of independent variables when the dependent variable is switched to "definite plans to stay" rather than "plans to stay."

The results are fairly consistent with those reported earlier for "plans to stay." For example, older individuals are less likely to stay. However, we cannot reject the hypothesis that marital status does not matter. We also find that in this instance the probability of staying is 0.024 lower for women than for men.

Experiences during doctoral training have powerful effects on the probability that a temporary resident has definite plans to stay in the United States. For example, having full-time employment during the last year of graduate school increases the probability of definitely staying by 0.26. Likewise, working part-time increases the probability by 0.18. The effect of having received one's bachelor's degree in the U.S. is similar, though slightly muted, compared to the case of planning to stay.

We find that country of origin plays less of a role in determining definite plans compared to plans to stay. Other things being equal, for example, being Chinese increases the probability by 0.23; being Indian increases the probability by 0.25. Being Brazilian or Mexican reduces the probability of definite stay plans by 0.2 or 0.1, respectively. Far from inconsequential, these effects are noticeably lower than the country effects in determining stay plans.

When the dependent variable is measured in terms of "definites," the field effects are considerably enhanced. Relative to the benchmark of biology (with its high postdoc rate), those with degrees in all other fields are less likely to definitely stay, and in many instances the effect is considerable. The field of training associated with the lowest likelihood relative to the benchmark is once again agriculture. Aerospace engineering, civil engineering, and oceanography are not far behind. Even those trained in the strong-demand sectors of electrical engineering and computer science report considerably lower definite plans to stay than those in biology. This effect washes out when we focus on those trained at top programs. Indeed, we conclude that temporary residents who earn degrees in computer science from top-rated programs are as likely to stay in the United States as are biologists.

A striking finding is that individuals trained at top programs are more likely to have definite plans to stay in the U.S. than are individuals who are not trained at top programs. There is but one exception: those trained at top medical programs are less likely to stay than are those trained at lower-rated medical programs.

We conclude that stay plans as well as definite plans are clearly related to age, field, country of origin, and quality of training. The fields with the highest stay rates are chemistry and biology. The field with the lowest stay rate is agriculture. Individuals trained at top programs consistently are more likely to have definite plans to stay than are individuals trained at lower-tier institutions.

Work and fellowship experiences in graduate school also clearly relate to staying. It is, of course, difficult to know whether these experiences are causal or reflect underlying characteristics that place these individuals in high demand in the United States. Suffice it to say, however, that these experiences are an extraordinarily good predictor of whether the individual will stay in the U.S.

Taken together, these effects can be quite strong. For example, the probability that a thirty-three-year-old unmarried male Chinese student supported on a fellowship who received a Ph.D. in 1997 from a top-ten-rated program in biology and did not earn a bachelor's degree in the United States plans to stay is 0.98; it is 0.76 that he has definite plans to stay. The probability for a comparable individual from India with a Ph.D. from a top institution in computer science is similar: 0.94 for plans to stay and 0.78 for definite plans to stay. For others the probability is much lower. For instance, the probability that a Mexican student who received a Ph.D. in agriculture from a non-top program plans to stay in the U.S. is 0.27; it is 0.15 that he has definite plans to stay.

Industrial Employment

For the years 1997–99 we not only know if individuals plan to stay in the United States, but, for those who have definite plans to work in industry, we know the identity and location of the firm where they have such plans (Stephan et al. 2004). We find that 32 percent of all Ph.D.s with definite plans to work in industry are temporary residents at the time of graduation. This is approximately the same proportion as the underlying representation of temporary residents in the population of new Ph.D.s during the time period. Definite plans to work in industry by field of training and resident status are provided in Table 6.4. The proportion of Ph.D.s going to industry who are temporary residents (column c) is particularly high in math (43 percent), civil engineering (42 percent), and electrical engineering (41 percent). The rate is also relatively high in mechanical engineering (40 percent) and computer science (38 percent). The proportion going to industry closely resembles the underlying proportion of temporary residents in the population of newly minted Ph.D.s during the same period (column d) for more than half the fields. Only in math and oceanography is the proportion greater than the proportion of all doctorates in these fields who are temporary residents; it is slightly higher in computer science. The proportion of Ph.D.s going to industry who are temporary residents is noticeably lower than the benchmark in agriculture, aeronautical engineering, chemistry, chemical engineering, and the biological sciences.

When we look at country of origin, we find that the largest number of foreign industrial hires among these new Ph.D.s comes from China. Indeed,

TABLE 6.4 Field of training of S&E temporary residents with definite plans to work in industry in the United States, 1997–99.

Field	Number of DIP* who are temporary residents (a)	Number of all DIP* (b)	Percent of all DIP* who are temporary residents (c=a/b)	Percent of all doctoral recipients who are temporary residents
Aerospace Engineering	36	159	22.6	33.1
Chemical Engineering	254	754	33.7	42.0
Civil Engineering	122	292	41.8	43.2
Electrical Engineering	763	1,860	41.0	40.4
Mechanical Engineering	293	738	39.7	40.8
Other Engineering	436	1,238	35.2	37.5
Agriculture	57	272	21.0	39.1
Astronomy	9	44	20.5	20.7
Biological Sciences	86	574	15.0	20.9
Chemistry	207	1,175	17.6	27.3
Computer Sciences	282	737	38.3	35.2
Earth Sciences	59	219	26.9	26.2
Mathematics	197	457	43.1	37.2
Medicine	73	415	17.7	17.8
Oceanography	s	s	33.3	26.6
Physics	181	626	28.9	32.9
All S&E	3,057	9,566	32.0	30.3

Note: s = suppressed if count is 6 or less.

* Doctoral industrial placements

the Chinese representation among the industrial hires is so substantial that almost one in three of the temporary residents hired by industry is Chinese and close to one in ten of all industrial hires identified in the data is Chinese. Indian hires are a close second, with more than one in twelve of industrial hires during the period being Indian. The prevalence of hires from China is consistent with the share of Chinese in the overall pool of new Ph.D.s during this period; the prevalence of hires from India is higher than the proportion of Indians in the Ph.D. population. The underlying proportion among new S&E Ph.D.s in 1997–99 is approximately 5 percent for Indians and almost 11 percent for Chinese.

A substantial portion of temporary residents going to industry are employed in large, established firms. Forty percent had definite plans for employment at a firm ranked in the top 200 for R&D expenditures or at one of these firms' subsidiaries. India placed the largest number of hires in a top-200 R&D firm, followed by China and Taiwan. These three countries accounted for 70 percent of temporary resident hires at a top-200 R&D firm; in comparison, they made up approximately two-thirds of temporary resident hires at non-top-R&D firms. The dominance of these countries among temporary residents going to industry no doubt reflects the substantial number of and high stay rates for new Ph.D.s from China, India, and Taiwan as well as their fields of study.

CONCLUSION

Temporary residents play a key role in S&E Ph.D. programs in the United States. During the period under study (1981–99), approximately one in three degrees in S&E was awarded to a student on a temporary visa. In certain fields, such as electrical engineering, math, and computer science, the percentage is significantly higher.

The lion's share of Ph.D. students on temporary visas during the past two decades came from four countries: the People's Republic of China, India, Taiwan, and South Korea. These patterns, however, are in the process of changing as Taiwanese and Korean students increasingly choose to remain in their country to receive their doctoral training.

Stay patterns have been increasing over time and are a major contributor to the internationalization of U.S. science. Stay patterns vary considerably by country of origin and field of training. Of the largest sending countries, the Chinese are most likely to stay; Brazilians are the least likely to stay. Biologists have the highest probability of staying; those trained in agriculture have the lowest probability of staying. These trends undoubtedly are influenced by the large number of postdoctoral positions available in the biological sciences. We also find strong evidence that those trained at top programs are more likely to plan to stay than are those trained at non-top programs.

The descriptive nature of this study leaves certain questions unanswered, inviting further research. For example, why have the number of temporary residents trained in top programs remained constant for most of the 1990s despite the fact that the overall number of temporary residents earning Ph.D.s steadily increased? Are there institutional factors, such as quotas on foreign students, that have driven this outcome? How have stay patterns been affected by the state of the U.S. economy? Did stay patterns decrease when the IT sector stumbled in 2000? How responsive are stay rates to conditions in the sending country? How successful are special initiatives in luring scientists and engineers back to their home country? A broader question is whether the changing composition of foreign students affects their education and employment patterns as well as those of native-born students. For instance, are underlying factors jointly influencing foreign students' decisions about studying in the United States, choosing a field, and staying after graduation? Is the high stay rate for foreign students in biology a factor in encouraging foreign students to choose to study biology in the U.S.? There is also the question of whether the high number of foreign-born students discourages U.S. students from entering careers in science and engineering. These—and other questions—invite further investigation.

Since the events of 9/11, U.S. policy toward foreign students has changed in a number of ways.[10] At the same time, the United States experienced slowed economic growth in the early 2000s, while the competition that started in the early 1990s for Ph.D. students from such countries as the United Kingdom, Germany, Japan, and China has continued. The role of foreign students in U.S. science is being affected by these and other factors. For example, first-year graduate enrollment data for the early 2000s indicates that the number of domestic students has increased while the number of international students has been flat or declined. Moreover, applications to graduate school by international students were down by 28 percent in 2003–4. Enrollments, however, were down by only 6 percent. Data for 2004–5 suggests that applications have continued to fall, but at a much slower pace than during 2003–4 (National Academy of Sciences 2005b). Foreign-student enrollments in 2004–5 dropped 7 percent from 2003–4, while domestic student enrollments fell 1 percent, the first drop since 2000.

A survey of U.S. educational institutions in the fall of 2003 found that 46 percent of responding institutions reported declines from the previous year in the number of newly enrolled international students. For over 40 percent of these institutions, new international enrollments fell by more than 10 percent (Institute of International Education 2004). Declines were more common than increases at all responding institutions for students from China, Saudi Arabia, Pakistan, the United Arab Emirates, Thailand, and Indonesia. New enrollments from Middle Eastern countries fell 10 percent between 2002 and 2003. The most common reasons cited by these institutions for declines in foreign students were new visa application processes, financial difficulties, and students selecting another host country for their education. Combined, these three reasons accounted for approximately 90 percent of the cited causes of declining new foreign enrollments.

While it is too early to know the degree to which these patterns may persist, the evidence reinforces the need for ongoing investigation of the role of temporary residents in U.S. education and science. From at least the early 1980s, doctoral education in science and engineering in the United States has become increasingly populated with and dependent on foreign students—as has the doctoral-trained U.S. workforce. If the number of foreign students studying (and staying) in the United States were to decline, there could be long-term implications for the health of the U.S. science and engineering enterprise.

NOTES

1. See www.nsf.gov/sbe/srs/ssed/start.htm.

2. Country of citizenship is not reported for 1.5 percent of the temporary residents receiving degrees during the decade.

3. In terms of population, China and India rank first and second, while South Korea ranks twenty-fifth and Taiwan forty-second.

4. A top program in a given field is defined as one ranked in the top ten based on the National Research Council's 1993 ranking of scholarly quality for all fields except agriculture and medicine (National Research Council 1995). A top program in these two fields is defined as being among the top ten institutions for federally funded R&D expenditures in the given field. For our fields that are more broadly defined than the NRC program definitions, such as biology, our rankings are based on the mean of all NRC-rated programs at an institution that fall under our field definition.

5. The apparent contradiction with the findings of Freeman, Jin, and Shen in chapter 10 (herein) reflects the fact that their analysis is for all foreign born, while

our analysis focuses especially on students who study on temporary visas. Moreover, their analysis measures quality in terms of the university, while our analysis examines quality at the individual program level.

6. Those who do not reply to the stay question tend to be slightly older, less likely to be married, and more likely to have received their Ph.D.s in the earlier period of observation.

7. The benchmark also includes "not employed" predoctoral status (PRENO-TEMP), male, unmarried individuals, and non-U.S. bachelor's degree.

8. Students study in the U.S. on an F-1 visa. Generally speaking, a student can stay up to one year after graduation on an F visa to obtain optional practical training. Most postdoc recipients hold J-1 visa status. The H visa is a temporary work visa and is specific for job and site. It is issued for up to three years and can be renewed for up to an additional three years. Some postdocs convert from a J-1 to an H-1B during their postdoctoral training.

9. Those from top programs do not always indicate a higher likelihood of staying. In particular, we find that those from top medical programs are less likely to stay than are those from non-top programs.

10. Under the auspices of Homeland Security Presidential Directive-2, the USA Patriot Act, and various immigration legislation, changes in immigration policy since September 11, 2001, that affect foreign students include: more stringent review of visa applications, implementation of the Student and Exchange Visitor Information System (SEVIS), development of the Interagency Panel on Advanced Science and Security (IPASS), clarification and increased usage of the MANTIS special visa review, creation of the new CONDOR special visa review, and implementation of the National Security Entry-Exit Registration System (Arnone 2004). For a summary of these and other changes, see National Academy of Sciences (2005b).

7

Do Foreign Students Crowd Out Native Students from Graduate Programs?

GEORGE J. BORJAS

INTRODUCTION

The Immigration and Nationality Act provides two types of "nonimmigrant" (i.e., nonpermanent) visas for persons wishing to study in the United States. The F-1 visa is for academic studies, and the M-1 visa is for vocational studies. The number of visas issued to foreign students increased greatly in recent decades. In 1980, 155,000 foreigners were granted temporary visas to study in the United States. By 2000 the number of student visas totaled 315,400, with the bulk of the visas (98 percent) granted to persons enrolled in academic programs.

As a result of the increasing number of visas granted to foreign students, the share of nonresident aliens enrolled in graduate programs in the United States rose from 5.5 percent in 1976 to 12.4 percent in 1999. As discussed in several chapters in this book, this increase had a particularly large impact on graduate enrollment in the sciences. By 1999–2000, nonresident aliens received 38.2 percent of all doctorates awarded in the physical sciences, 52.1 percent in engineering, 26.6 percent in the life sciences, and 22.8 percent in the social sciences (National Center for Education Statistics 2002, tables 208, 275).

Remarkably, there has been practically no research analyzing the costs and benefits of foreign students. We know almost nothing about their impact on the higher education system, on the U.S. labor market, and on the economies of the source countries.[1] Instead, most of the work that has been done focuses on foreign-born scientists and engineers more generally, such as the chapter by Stephan and Levin (herein).

This study investigates how the enrollment of domestic students relates to the rapid growth in the number of foreign students enrolled in graduate school. If the number of slots in graduate programs was fixed, and qualified natives were applying to programs, an increase in the number of foreign students would imply that natives, who would presumably have filled those positions, have been crowded out. Even if the graduate programs were expanding, an increase in the supply of foreign students might sufficiently alter incentives for natives to pursue some programs, particularly if many foreign students stay in the United States and reduce economic opportunities in some occupations.

On aggregate, the empirical analysis reported in this paper suggests that there is little evidence of a crowd-out effect for the typical native. This result, however, masks sizable differences in the impact of foreign students across ethnic groups and between native men and women. In fact, there is a strong negative correlation between increases in the number of foreign students enrolled *at a particular university* and the number of white native men enrolled in that university's graduate program. The study thus suggests that the growth in foreign student enrollment may have altered the educational opportunities available to white native men.

THE CROWD-OUT EFFECT

A department's decision to admit additional foreign students to its graduate program obviously depends on many factors, including the relative quality of the applicants, the possibility that foreign students pay for a higher fraction of their education (either directly or through fellowships that they bring with them), the widespread adoption of the axiom that "diversity" is beneficial in a university setting, and the relative marginal products that result from having foreign and native students as employees of the university (since many graduate students typically work as research assistants or teaching assistants). Some of these factors may imply that, other things being equal, faculty would prefer to admit a foreign student instead of a native-born applicant.[2]

The admission and eventual enrollment of foreign students alters the educational opportunities available to qualified natives in two distinct ways. First, it may be the case that the number of slots available in a particular graduate program is fixed in the short run. The enrollment of an additional foreign student would then necessarily imply that one fewer native student would be enrolled, assuming that native students continue to apply to the program. This is the simplest and clearest case of what we call a crowd-out effect. Because quality is but one of several attributes that may be considered in making admissions decisions, the presence of such an effect says nothing about the quality of the marginal student. To make quality inferences, we would have to know how the foreign students compare to the denied native student.

Even if the university were expanding and admitting more foreign *and* more native students, there may still be a crowd-out effect in the sense that native enrollment would have risen faster if the university had not increased its supply of foreign students. In the empirical analysis reported below, I adopt the conservative definition of a crowd-out effect that requires native enrollment to actually fall (rather than not rise as much as it would have risen otherwise) when the number of foreign students increases.

The entry of foreign students can alter the educational decisions made by native students in another, less direct, way. In particular, an increase in the number of enrolled foreign students may affect the incentives that natives have to pursue some educational programs. Suppose, for instance, that many of the foreign students enrolled in a particular program (e.g., computer science) remain in the United States after graduation. One would then expect that wages in these computer-related occupations would fall and those occupations would become relatively less attractive to natives.[3] The foreign students may still choose to enter those low-paying jobs if their career decisions are mainly guided by the fact that the student visa is perceived as providing an entry ticket into the United States, so that they would be comparing the low U.S. wage in a computer-related occupation with the even lower wage that would be available if they remained in the source countries. In contrast, native students have many more career choices and would shy away from applying to those educational programs where foreign students cluster. In the long run, this behavioral response would again imply that an increase in the enrollment of foreign students in a particular program would reduce the number of natives enrolled in that program.[4]

There is one important distinction between the two types of crowd-out effects discussed above. The first crowd-out effect is specific to a particular university—and indicates how native enrollment in that institution changes as the number of foreign students enrolled in that institution increases. The second crowd-out effect results from an economy-wide behavioral response that effectively inhibits natives from pursuing particular educational programs in all universities (or perhaps from pursuing a graduate education altogether if the labor supply increase resulting from the foreign student program is sufficiently large in all fields). The empirical analysis presented below shows these economy-wide fluctuations and examines the shifts that occur in native enrollment *within a particular university* as the size of the foreign student population increases. The study, therefore, will isolate the institution-specific type of crowd-out effect.

DATA

Since 1986, the Integrated Postsecondary Education Data System (IPEDS) has collected detailed information on enrollment, employment, and finances in institutions of higher education. Each institution reports the number of persons enrolled in particular programs both at the undergraduate and graduate levels, including the gender and race of students as well as the number of nonresident aliens (which, for simplicity, I will refer to as "foreign students"). The educational institution also reports detailed information on expenditures in various categories relevant to the higher education sector (e.g., instruction and research). Prior to 1986, the same type of information was collected by the Higher Education General Information System (HEGIS), a precursor of the IPEDS data. My empirical analysis uses enrollment information provided by both of these surveys.

My analysis focuses on enrollment trends in graduate programs. These enrollment statistics do not include students who attend professional schools in law, medicine, or dentistry. They do include students enrolled in graduate business programs, such as the MBA, as well as students enrolled in master's programs in education, humanities, social sciences, and sciences and engineering. At the doctoral level they include students enrolled in sciences, engineering, social sciences, education, humanities, and "other." As Groen and Rizzo show in Table 9.1, the majority of these doctoral degrees are now bestowed in the nonsciences.[5] My analysis is further restricted to higher education institutions in the United States that are accredited at the college level by the U.S. Department of Education and that are legally authorized to offer at least a one-year program of study creditable to a degree.[6] My empirical study of enrollment trends uses the cross-sections observed in 1978, 1982, 1986, 1990, 1994, and 1998. (The multivariate regression technique used in the analysis is discussed in the statistical section of this volume.)

In each of these cross-sections, I calculate the total number of graduate students enrolled in each institution, regardless of whether they are enrolled full time or part time. The choice of the timing of the cross-section snapshots is due to two factors. Prior to the 1990s, the IPEDS surveys were not conducted annually, and some of the available cross-sections do not contain any information on the number of foreign students enrolled in the institution. Further, the four-year gap across cross-sections implies that there is a significant turnover in the graduate student population of a particular institution from survey to survey, minimizing the problems that would arise if many students were double-counted because they appeared in several surveys.[7] Because the IPEDS contains only limited information on field of study, I restrict the analysis to the size of the entire graduate program at a particular institution.

Table 7.1 summarizes the enrollment trends for various groups of students. The number of foreign students more than doubled between 1978 and 1998, from 79,400 to 194,300. There has also been a sizable increase in the number of native-born graduate students. In 1978, there were 1.2 million native graduate students, and this number increased to 1.4 million in 1990 and to 1.6 million in 1998. It turns out, however, that *all* of this growth

TABLE 7.1 Enrollment trends in graduate programs, 1978–98.

Group	Number of students (in 1000s) by year					
	1978	1982	1986	1990	1994	1998
Nonresident aliens	79.4	105.0	132.4	167.3	179.5	194.3
Male	60.7	78.7	97.0	116.4	118.1	120.8
Female	18.7	26.3	35.4	50.9	61.4	73.4
All natives	1,239.3	1,217.3	1,302.9	1,418.8	1,542.0	1,569.6
Male	627.0	591.0	596.3	621.0	657.7	631.3
Female	612.3	626.3	706.6	797.9	884.3	938.2
Asian natives	27.5	35.0	41.7	53.2	72.6	86.2
Male	16.3	20.9	24.5	29.7	38.3	41.8
Female	11.1	14.2	17.2	23.6	34.3	44.4
Black natives	76.4	68.9	70.3	83.9	110.6	138.6
Male	29.9	26.1	25.6	29.3	37.7	44.1
Female	46.5	42.8	44.7	54.6	72.9	94.5
Hispanic natives	27.9	31.7	44.4	47.2	63.9	82.7
Male	14.4	14.8	19.9	20.6	27.0	32.4
Female	13.5	17.0	24.6	26.6	36.9	50.3
White natives	1,094.0	1,074.7	1,101.4	1,228.4	1,286.8	1,252.4
Male	556.0	525.5	505.2	538.8	551.4	509.3
Female	537.9	549.2	596.2	689.5	735.4	743.1

Source: Statistics calculated using the HEGIS (pre-1982) and the IPEDS (post-1986) data files.

occurred among native women. In contrast, the number of male native-born students hovered around 600,000 throughout the entire period.

One particular group of natives—white men—will play a significant role in the analysis reported below. They comprise the only native group that had a lower enrollment in graduate programs at the end of the period than at the beginning. In particular, there were 556,000 white native men enrolled in graduate programs in 1978. This statistic fell to 539,000 in 1990 and to 509,000 in 1998. Note, however, that graduate enrollment for this group did not decline monotonically throughout the period and was almost at its 1978 level in 1994.

ESTIMATING CROWD-OUT EFFECTS

Let N_{it} denote the number of native graduate students enrolled in university i at time t, and let F_{it} denote the respective number of foreign students. Much of the statistical evidence reported in this paper is obtained by stacking the enrollment data obtained from the HEGIS and IPEDS across universities and surveys and estimating the regression model

(1) $\quad N_{it} = \Theta \, F_{it} + s_i + \pi_t + \varepsilon_{it}$

where s_i is a vector of fixed effects indicating the university and π_t is a vector of fixed effects indicating the time period. The university fixed effects absorb any university-specific factors that may determine the size of native enrollment. The period fixed effects absorb any time-specific factors that determine the size of the native population interested in pursuing a graduate education at a particular point in time. Throughout the analysis, the regression will be weighted by the total enrollment of the graduate program in a particular university at a particular point in time (or $N_{it} + F_{it}$). Further, the standard errors are clustered by university to adjust for possible serial correlation within a particular institution.

Under some conditions, the magnitude of the coefficient Θ provides information about what we have termed the crowd-out effect suggested by the enrollment data. In particular, Θ measures what happens to native enrollment *within a particular university* when that institution enrolls one more foreign student. If the estimate of Θ were zero, for example, the data would indicate that the enrollment of an additional foreign student simply expands the size of the university and has no effect on its preexisting (native) enrollment. If the estimate of Θ were -1, there would be a one-to-one trade-off between natives and nonnatives. The total number of students enrolled in the university's graduate program is constant, and each additional foreign student displaces a native student who presumably would have otherwise enrolled. Of course, Θ may also be positive, perhaps even exceeding 1. Over time, some universities have expanded, and the coefficient Θ measures how this expansion affected the relative enrollment of native and foreign students.

The top panel of Table 7.2 reports the coefficient Θ estimated from various specifications of the model. Each coefficient reported in the table is estimated from a different regression model, where the dependent variable is the number of native graduate students in a particular race-gender group. Consider the first regression coefficient reported in the table, where the dependent variable is the total number of natives (both men and women) enrolled in school i at time t and the independent variable gives the total number of foreign students enrolled in that school at that time. The estimated coefficient is 0.046 (with a standard error of 0.279), indicating that an additional foreign student, at the margin, had no impact on the number of natives enrolled at that institution.

This aggregate correlation, however, masks a great deal of dispersion, particularly in terms of the impact of foreign students on the enrollment of natives who differ in their gender and ethnic background. Most important, Table 7.2 documents the existence of a significant negative correlation

TABLE 7.2 Impact of foreign students on native enrollment.

Institutions	Ethnicity	Sex		
		Male and female	Male	Female
All schools	All natives	0.046 (0.279)	−0.198 (0.152)	0.244 (0.141)
8,236	Asian natives	0.232 (0.054)	0.105 (0.025)	0.127 (0.030)
	Black natives	0.105 (0.026)	0.033 (0.009)	0.071 (0.019)
	Hispanic natives	0.191 (0.126)	0.080 (0.054)	0.111 (0.073)
	White natives	−0.488 (0.268)	−0.418 (0.139)	−0.070 (0.145)
Public institutions	All natives	0.214 (0.342)	−0.093 (0.178)	0.307 (0.177)
3,103	White natives	−0.197 (0.259)	−0.272 (0.139)	0.075 (0.135)
Private institutions	All natives	−0.194 (0.404)	−0.328 (0.227)	0.134 (0.208)
5,133	White natives	−0.856 (0.428)	−0.589 (0.222)	−0.267 (0.239)

Notes:
Numbers of observations are in italics.
Standard errors are reported in parentheses and are clustered by institution.
All regressions include a vector of fixed effects indicating the institution and a vector of fixed effects indicating the survey year.

between foreign students and the enrollment of white native men. For this group, the coefficient is -0.418 (0.139). This negative coefficient does not indicate that graduate enrollment for this group was declining at every university. That potential trend is absorbed by the period fixed effects included in the regression model. Instead, the estimated coefficient indicates that the enrollment of white native men fell most in those schools that had larger increases in the number of foreign students enrolled. In short, the evidence suggests a significant *institution-specific* negative correlation between the enrollment of foreign students and white men.[8]

The raw data underlying this result can be easily illustrated. Figure 7.1 presents the scatter diagram that relates the 1978–98 change in the enrollment of white native men to the respective change in the number of foreign students enrolled at a particular university. Each point in the scatter diagram, therefore, represents enrollment changes that occurred at a school over the period. It is clear that the enrollment of white native men fell most steeply in those schools that had the largest increases in foreign student enrollment.

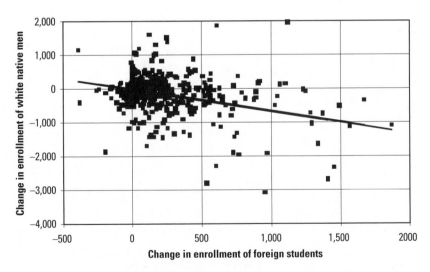

FIGURE 7.1 Change in enrollment of foreign students and white native men, 1978–98.

Notes:
Each point in the scatter diagram indicates the 1978–98 change in foreign students and white native men for a particular university.
The regression line weighs the data by the total graduate enrollment at the university (as of 1998). The coefficient is -0.649, with a standard error of 0.053.

Because graduate education for both foreign and native students is highly subsidized by U.S. taxpayers (Winston 1999), it is important to determine if the crowd-out effect differs between public and private institutions. The bottom two panels of Table 7.2 report some of the regression coefficients estimated in each of the two sectors. Although the crowd-out effect on white native men is negative and significant in both sectors, it is substantially larger in private universities. Moreover, the regression for white native *women* suggests that they may be crowded out of private institutions as well. In fact, the coefficient for total white native enrollment in the private sector is -0.856 (0.428), suggesting a one-to-one displacement of white natives as foreign student enrollment increases.[9]

CROWD-OUT AND SCHOOL QUALITY

Any policy evaluation of a potential crowd-out effect will depend on the constraints that the enrollment of foreign students imposes on the educational access available to natives. As a result, it is important to examine how

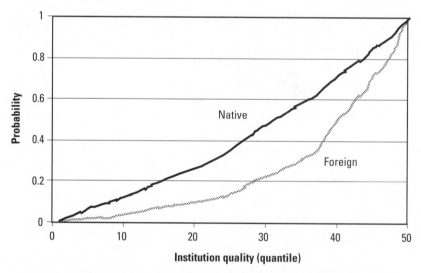

FIGURE 7.2 Cumulative probability distributions for native and foreign graduate enrollment, 1998.

Source: Enrollment data drawn from the 1998 IPEDS.

Note:

The quality ranking for an institution is based on the institution's per-student instructional expenditure between 1990 and 1993; see the text for more details.

the crowd-out effect varies across institutions that differ in the quality of their graduate programs.

To assign a quality ranking to a particular institution, I used the data on instructional expenditures reported in the IPEDS files. I calculated the *average* per-student instructional expenditures for the survey years 1990–93.[10] The averaging of the expenditure data over the four years helps to minimize the problem of both measurement error and short-run fluctuations in instructional expenditures. I divided the population of institutions into fifty quantiles. There are approximately 1,100 institutions in my data extract, so each quantile of the distribution contains around twenty-two schools. The top two quantiles of this distribution contain the list of "usual suspects," including Harvard, Yale, Princeton, Columbia, and the California Institute of Technology.

Figure 7.2 presents the cumulative probability distributions for the stock of foreign and native students (as of 1998) along the quality spectrum.

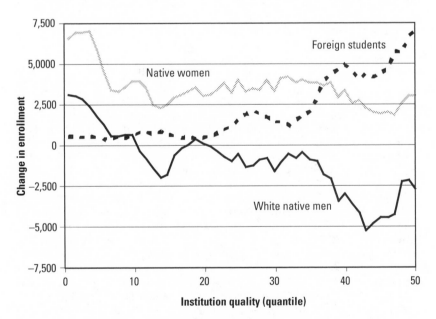

FIGURE 7.3 Change in enrollment between 1978 and 1998 by quality of institution.

Source: Enrollment data for each institution drawn from the 1978 HEGIS and the 1998 IPEDS.

Notes:
The quality ranking for an institution is based on the institution's per-student instructional expenditure between 1990 and 1993; see the text for more details.
The lines represent a five-quantile moving average.

Native enrollment is much more evenly distributed across the spectrum than is foreign enrollment. For instance, 48 percent of natives are enrolled in schools in the bottom thirty quantiles of the quality distribution, as compared to only 22.8 percent of foreign students.

There is also an important difference in how enrollment grew for foreign and native students across the various quantiles. Figure 7.3 shows that native enrollment between 1978 and 1998 grew fastest at the lower-quality institutions, while foreign enrollment grew fastest at the higher-quality institutions. The figure also suggests that the crowd-out effect of foreign students on white native men isolates a unique relationship that is not found when one contrasts the enrollment trends of white native men with other groups. The 1980s and 1990s witnessed not only a large increase in foreign enrollment, but also a large increase in the number of native women enrolled in graduate programs. Figure 7.3 shows that native women are much more evenly scattered across the quality spectrum, so the growth of women

TABLE 7.3 Impact of foreign students on native enrollment by quality of institution.

Native group	Quantile of quality distribution							
	1st–10th *1,101*		11th–40th *3,632*		41st–50th *1,216*		50th *115*	
All natives	9.358	(3.371)	0.515	(0.323)	−0.176	(0.294)	−0.466	(0.558)
Men	3.911	(1.668)	0.104	(0.152)	−0.308	(0.166)	−0.466	(0.291)
Women	5.448	(1.745)	0.411	(0.192)	0.132	(0.153)	0.000	(0.268)
All Asians	0.169	(0.105)	0.151	(0.026)	0.216	(0.091)	0.165	(0.034)
Men	0.097	(0.041)	0.079	(0.018)	0.088	(0.043)	0.051	(0.023)
Women	0.072	(0.066)	0.072	(0.009)	0.128	(0.050)	0.114	(0.013)
All blacks	0.377	(0.170)	0.121	(0.030)	0.117	(0.038)	0.149	(0.041)
Men	0.200	(0.068)	0.044	(0.011)	0.033	(0.012)	0.045	(0.014)
Women	0.177	(0.150)	0.077	(0.021)	0.084	(0.028)	0.104	(0.028)
All Hispanics	8.090	(4.171)	0.074	(0.041)	0.094	(0.037)	0.083	(0.007)
Men	3.427	(1.805)	0.034	(0.015)	0.033	(0.011)	0.027	(0.004)
Women	4.663	(2.367)	0.040	(0.027)	0.061	(0.029)	0.056	(0.004)
All whites	0.719	(1.035)	0.191	(0.308)	−0.649	(0.322)	−0.882	(0.488)
Men	0.188	(0.281)	−0.038	(0.140)	−0.493	(0.167)	−0.605	(0.253)
Women	0.530	(0.801)	0.229	(0.187)	−0.157	(0.179)	−0.277	(0.236)

Notes:
Numbers of observations are in italics.
Standard errors are reported in parentheses and are clustered by institution.
All regressions include a vector of fixed effects indicating the institution and a vector of fixed effects indicating the survey year.

occurred in the same types of institutions where the enrollment of white native men grew most (or decreased least). In short, there is no evidence that the increase in the supply of native women in graduate programs crowded out native white men.

Table 7.3 reports the coefficient Θ from regression models estimated separately in sets of institutions of roughly similar quality. The correlation between the enrollment of white native men and foreign students is slightly positive for the lower-quality schools (0.188, with a standard error of 0.281), turns zero for schools in the middle of the quality distribution (-0.038, with a standard error of 0.140), and becomes negative for schools at the top of the distribution (-0.493, with a standard error of 0.167). The coefficient is most negative when the regression model is estimated in the subset of elite institutions in the fiftieth quantile. The crowd-out effect for white native men is then -0.605 (0.253). In fact, the table suggests that there may also be a crowd-out effect for white native women at these elite institutions.

REVERSE CAUSALITY

An alternative interpretation of the negative sign of the coefficient Θ is that universities increased their enrollment of foreign graduate students because they faced a substantial decline in the number of (qualified) white native men who wished to enroll in their graduate programs. This is consistent with the work of Attiyeh and Attiyeh (1997), who examine factors influencing graduate admissions decisions at forty-eight graduate schools in five disciplines. Their analysis shows that in four out of the five fields, graduate schools in the aggregate gave substantial preference to U.S. citizens over foreign applicants.

If fewer white men are applying, one must ask why. An obvious reason leading to a decline in the number of white native men could be changing demographics: there may have been a substantial decline in the size of the population of white native men who are college graduates. One problem with this hypothesis is that the period fixed effects included in the regression model already control for such demographic trends. Moreover, the actual demographic trends are *not* entirely consistent with this hypothesis The pool of potential graduate students among white native men rose dramatically between 1970 and 1980 (as the baby boomers reached their twenties) but declined by only about 200,000 persons since then, with almost all of the decline occurring between 1980 and 1990. In contrast, the number of native white men enrolled in graduate programs has hovered between 500,000 and 550,000 throughout the past two decades.[11]

Alternatively, one could argue that white native men simply no longer wish to pursue careers that require a graduate education. In the terms of Groen and Rizzo (chapter 9 herein), the propensity of white men to attend graduate school has declined. This is hard to refute, because our data are aggregated across all graduate programs within a university, while Groen and Rizzo examine propensities by field of training. For example, they find that the propensity to get an M.B.A. has increased for domestic men while the propensity to get a Ph.D. has remained stable since the early 1980s, after declining significantly in the 1970s.

The argument that white native men simply no longer wish to pursue careers that require a graduate education resembles the often-heard argument that "immigrants do jobs that natives do not want to do" in the U.S. labor market. If one takes this set of arguments seriously, it would seem to imply that there are few jobs that natives *do* want to do. The argument is used to justify why immigrants do not compete with low-skill workers, such as gardeners and taxi drivers—because natives do not want to perform those presumably menial jobs. It is also used to justify why increased immigration of high-tech workers is needed to alleviate presumed labor shortages in the high-tech sector—because natives do not want to be software programmers. And, in this context, it would be used to explain why immigrants do not crowd out natives from graduate programs—because white native men no longer wish to attend graduate school.

However, it is unlikely that this is a valid conjecture. After all, the greatest declines in the enrollment of white native men occurred at high-quality institutions. Surely all potential graduate students—regardless of their race—would prefer to attend those institutions that provide the best job opportunities after graduation. Of course, it is possible that it is the *most* qualified white native men who no longer wish to attend graduate school, so that graduate programs have a shortage of qualified applicants and are forced to fill in the existing slots with foreign applicants. Moreover, note that the number of white native *women* attending elite institutions was adversely affected by the growing enrollment of foreign students, despite the very large increase in the number of women enrolling in graduate programs.

A more nuanced analysis would ask if the findings relate to the considerable heterogeneity that exists across graduate programs and types of schools. By way of example, although more than 1,100 colleges and universities grant graduate degrees in the United States, in 2000 only 339 granted doctoral degrees in science and engineering, and this number has grown considerably in recent years (Freeman et al., chapter 10 herein). The finding that native enrollment has grown in lower quantile programs may relate to the tremendous increase that has occurred in enrollment in non-Ph.D.-granting institutions and the types of graduate degrees awarded by these institutions. This is consistent with the finding that native men and women have similar quantile "quality" profiles. It also would explain why there is no indication of displacement at these institutions since noncitizens will be less likely to enroll in these programs, especially programs in education and the humanities.

Conclusion

The evidence presented in this paper documents a strong negative correlation between the enrollment of white native men in graduate programs and the enrollment of foreign students. Those educational institutions that experienced the largest increases in foreign enrollment are also the institutions that experienced the steepest drops in the enrollment of white native men. The evidence is consistent with the hypothesis that foreign students limit the opportunities available to white native men in graduate education, particularly at the most elite institutions. It is also consistent with the alternative hypothesis that graduate programs, in an effort to maintain their program size, increase the admission of foreign students in the face of declining applications from white men. To differentiate between the two requires application data at the department level. To the best of our knowledge, Attiyeh and Attiyeh (1997) are the only ones to do this, and they only do this for a limited number of programs.

It is worth stressing that if it is the former, the potential crowd-out effect of foreign students on the enrollment of white native men may not signal a suboptimal allocation of resources in the graduate education sector. The implications of the finding depend crucially on three related issues. First, what happens to the displaced white native men? Second, what happens to the foreign students after they complete their education? Finally, what are the costs and benefits that foreign students impart on universities and on the U.S. economy?

The first of these questions is difficult to answer, as we simply do not know the career choices made by the displaced white men (and it is difficult to imagine a simple way of measuring this counterfactual). Did these men move on to lower-quality graduate programs, or did many of them decide to forego a graduate education altogether? Moreover, any cost-benefit analysis requires information on how these men would have fared had they attended the graduate program of their choice.

We could potentially learn more about the career choices and contributions made by the foreign students after they complete their graduate education. Over 50 percent of foreign students who received their doctorates in the 1990s stayed in the United States (Finn 2000). What is the nature of the selection that determines the population of stayers?

Finally, foreign students impart many other costs and benefits. They pay tuition, and these tuition revenues—if they were to exceed the actual cost of providing an education—could be an additional source of economic benefit. But the pricing of higher education in the United States is highly distorted in both private and public institutions, with the typical tuition payment not being sufficiently large to cover the actual cost of an education. If immigration policy is supposed to benefit the native population, it may be difficult to justify a subsidy system that limits educational opportunities for many native students unless the economic gains from foreign students are very large. Although we do not know the size of these gains, it is well documented that foreign doctorates who reside in the United States contribute disproportionately to the advancement of science (Stephan and Levin 2001). At the same time, however, existing calculations of the economic benefits from immigration tend to suggest that the net benefit that accrues to the native-born population is small (Borjas 1995; Johnson 1998).

In sum, although the rapid growth of the foreign student program in the past two decades has altered the higher education sector in significant ways, there is still much to learn before we can objectively assess the costs and benefits of this important shift in the parameters of U.S. immigration policy.

NOTES

1. A few studies examine how foreign-born teaching assistants affect the educational outcomes experienced by native-born undergraduates; see, for instance, Jacobs and Friedman (1988); Borjas (2000); and Fleisher, Hashimoto, and Weinberg (2003). Hoxby (1998) and Betts (1998) present the only studies that explicitly focus on measuring the impact of immigrants on the high school completion rates or enrollment rates in undergraduate programs of native students. Borjas (2002) summarizes some of the available evidence and details the research questions that would lie at the core of any cost-benefit evaluation of the foreign student program.

2. For example, there is anecdotal evidence (in cartoons in scientific newsletters

and so on) regarding the desirability of international students who will keep their noses to the grindstone.

3. Borjas (2003) presents the most recent evidence on the wage impact of immigration in the U.S. labor market. He finds that an immigration-induced 10 percent increase in the number of workers belonging to a particular skill group lowers the wage of that group by 3 to 4 percent.

4. Freeman et al. (2001) describe how the major shifts in the bioscience job market, shifts partly due to the influx of foreign students, alter the incentives for pursuing careers in that sector.

5. In the IPEDS file, professional students include students enrolled in professional programs in schools of law, medicine, and dentistry but do not include students attending business or engineering schools. The business and engineering students are classified as graduate students and are therefore included in the analysis that follows.

6. Operationally, this sample restriction limits the study to institutions that have a valid Federal Interagency Committee on Education (FICE) code.

7. Alternatively, I could have analyzed enrollment trends for first-time graduate students (an enrollment statistic that is also reported in the IPEDS). These data, however, seem to contain significant measurement errors, particularly in the earlier surveys.

8. The estimated crowd-out effects would be even stronger if the enrollment data also included enrollment in professional programs. In particular, the estimated coefficient for native white men would be -0.576 (0.145).

9. It is worth stressing that the potential crowd-out effect of foreign students on white native men isolates a unique relationship that is not found when one contrasts the enrollment trends of white native men with other groups. A regression of the enrollment of white native men on the number of foreign students and the number of white native women (as well as institution and period fixed effects) indicates that an increase in the number of foreign students reduces the enrollment of white native men, but an increase in the number of white native women does not. The estimated coefficients are -0.539 (0.109) and 0.496 (0.060), respectively.

10. The denominator includes all undergraduate and graduate students, regardless of whether they are enrolled part time or full time. The expenditure data was deflated using the CPI-U series. I restricted the set of institutions to those that reported an average per-student expenditure of less than $100,000 (eliminating mainly medical, law, and theological schools from the data). The data on instructional expenditures is not available for approximately 20 percent of the institutions. These institutions are omitted from the analysis reported in this subsection.

11. See also the related evidence presented by Groen and Rizzo (2003).

8

Foreign Scholars in U.S. Science:
Contributions and Costs

PAULA E. STEPHAN AND SHARON G. LEVIN

INTRODUCTION

The foreign born have an exceptionally strong presence in U.S. science. They make up a large and increasing percentage of the scientific workforce, and an increasing percentage of Ph.D. degrees in science and engineering are awarded to foreign-born students (see chapters 6 and 10 herein).

The presence of the foreign born in U.S. science raises several policy questions. One is the question of contribution: Do the foreign born and foreign educated contribute disproportionately to U.S. science? Another is the question of incidence: How does the increased supply of foreign-born scientists and engineers working in the U.S. affect the labor market outcomes of native-born scientists and engineers? A related question is whether the presence of foreign-born scientists and engineers discourages U.S. citizens from choosing careers in science and engineering.

Here we focus on the contribution and incidence questions, noting that the issue of contribution and incidence resembles what economists refer to as a welfare problem, much like the case of free trade, where the overall economy can benefit from free trade but individuals or groups of individuals incur substantial costs. In this chapter we summarize overall trends in terms of the workforce, focusing on the changing composition of the foreign born, as well as the changing composition of those studying on temporary visas. We then address the question of contribution, examining the birth and educational origins of individuals making significant contributions to U.S. science. Using a novel adaptation of the shift-share technique, we examine how the heavy inflow of foreign talent receiving doctorates in the United States is related to science and engineering (S&E) jobs held by citizens,[1] especially the choice positions within the academic sector, over the period 1979–97.

THE INCREASING PRESENCE OF THE FOREIGN BORN

The birth and educational origin of the U.S. scientific workforce can be examined using two related, but not strictly comparable, databases. The first, known as the National Survey of College Graduates (NSCG), is based on the U.S. census and has the virtue of identifying individuals working in the United States but trained outside the United States as well as individuals trained in the United States. The NSCG's drawback is that it is seldom fielded and thus leaves large gaps in our knowledge concerning the workforce. The second, known as the Survey of Doctorate Recipients (SDR), has the advantage of being fielded every other year. The drawback is that it only examines scientists and engineers working in the United States who received their doctoral training in the United States.

Table 8.1 presents data concerning the birth and educational origins of scientists and engineers working in the United States in 1980 and 1990 using the NSCG and thus including scientists whose doctoral training was received outside the United States.[2]

Regrettably, comparable data is not yet available for 2000. We exclude from the analysis individuals not in the labor force, individuals in the military, individuals not in the United States, and individuals in social science occupations. We restrict our definition of highly trained scientists to those who have a doctoral or medical degree and of highly trained engineers to those who have a baccalaureate degree. We use the NSCG to determine the size of the scientific workforce in 1990 as well as in 1980. For the latter, we restrict the sample to those who immigrated to the United States or completed their highest degree before 1980.[3]

Distributions are presented in Table 8.1 for five fields: engineering, physical sciences, mathematical and computer sciences, earth and environmental sciences, and life sciences. We see that 18.3 percent of the highly skilled scientists in the United States in 1980 were foreign born. The percent was highest among physical scientists (20.4 percent) and the lowest among life scientists (15.4 percent). By 1990 the proportion of foreign born had increased to 24.7 percent. More than one in four physical scientists and math and computer scientists working in the United States had been born abroad; for life scientists the proportion had increased from approximately one in seven to one in five. The proportion of engineers who are foreign born is

TABLE 8.1 Birth and educational origins of the scientific labor force in the United States, 1980 and 1990.

	1980				1990			
		Percent foreign				Percent foreign		
Field	Total	Born‡	Bacc.	Ph.D.	Total	Born‡	Bacc.	Ph.D.
All sciences*	55,697	18.3	13.6	8.8	120,888	24.7	16.0	10.7
Earth & environmental	4,048	17.6	12.3	19.0	6,976	17.4	9.6	13.5
Life	14,890	15.4	12.2	9.4	37,717	21.7	12.8	11.6
Math & computer	13,149	18.4	13.9	7.2	31,916	28.5	18.1	7.9
Physical	23,610	20.4	14.5	7.5	44,279	25.6	18.1	11.6
Engineering†	602,722	13.9	7.4	§	1,108,367	15.9	7.4	§

Source: Estimated from the 1993 NSCG (see text).

Note: Bacc. = Baccalaureate degree; Ph.D. = doctoral/medical degree

* Excludes individuals without doctoral or medical degrees, those not in the labor force, those not in the United States, those in the military, and those in engineering or social science occupations.
† Excludes individuals without a baccalaureate degree, those not in the labor force, those not in the United States, and those in the military.
‡ Includes individuals born abroad to U.S. citizens who are classified as "immigrants" in the NSCG.
§ Professional engineers often do not have doctoral degrees.

substantially smaller than that of highly trained scientists. In 1980 approximately 14 percent were foreign born; this had crept up to about 16 percent by 1990.

The disparate rates of growth in the native- and foreign-born components of the scientific labor force can be seen from Figure 8.1. In computer sciences the rate of growth of the foreign born was more than twice that of native born; in the life sciences it was approximately twice as great. Only in earth and environmental sciences has the rate of growth been about the same.

Many immigrants come to the United States to receive training and subsequently stay to work (Riess and Thurgood 1991; Finn 1997; Black and Stephan herein). Some come prior to receiving their undergraduate degrees, others afterward. Of the former, many migrated with their families when they were children. Another striking feature is the large number who come to the United States after receiving their doctoral training. In all but mathematics and computer science, more than one out of ten individuals in the U.S. scientific workforce in 1990 received their doctoral training abroad.[4]

Table 8.2 takes a longer, albeit edited, view of the presence of the foreign born in the workforce, using the SDR. We see over the period 1973–97

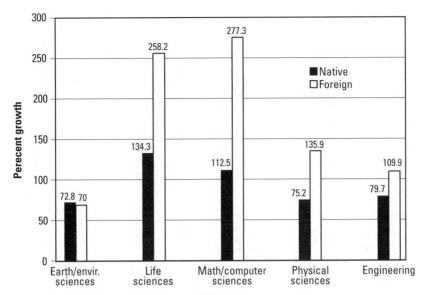

FIGURE 8.1 Growth in the native- and foreign-born components of the highly trained scientific labor force in the United States, 1980–90.

Source: Data from NSF, National Survey of College Graduates, 1993.

that the number of U.S.-trained doctoral scientists and engineers living in the United States who were citizens (either native born or naturalized) at the time their degree was awarded increased almost three-fold; those who held temporary or permanent visas at the time the degree was received increased eight-fold. Thus, while fewer than one in ten U.S. doctoral-trained scientists working in the United States in the early 1970s were not citizens at the time their degree was awarded, by 1997 more than one in five were not citizens. The citizen-noncitizen growth differential is most striking in the mathematical and computer sciences, followed by engineering and earth/environmental sciences. In the former, the number of citizens grew almost three-fold while the number of noncitizens grew thirteen-fold; in the latter two, citizen growth was more than two-fold, while non-citizen growth was more than nine-fold.

THE QUESTION OF CONTRIBUTION

We examine whether the foreign born and foreign educated contribute disproportionately to U.S. science by testing whether the foreign born and

TABLE 8.2 Growth in science and engineering (S&E) doctorates by field of training and citizenship status at the time the degree was earned in the United States.

Field	All S&E doctorates			U.S. citizen doctorates			Non-citizen doctorates		
	1973	1997	Growth	1973	1997	Growth	1973	1997	Growth
All fields	110,914	367,617	231.4%	101,506	290,980	186.7%	9,408	76,637	714.6%
ENG	26,649	87,585	228.7%	23,220	56,426	143.0%	3,429	31,159	808.7%
LIFE	36,050	142,330	294.8%	33,668	123,386	266.5%	2,382	18,944	695.3%
BIO	25,951	105,842	307.9%	24,342	91,882	277.5%	1,609	13,960	767.6%
PHYS SCI	48,215	137,702	185.6%	44,618	111,168	149.2%	3,597	26,534	637.7%
EAR/ENV	4,621	15,916	244.4%	4,397	13,896	216.0%	224	2,020	801.8%
CHEM	20,567	54,327	164.1%	18,936	44,968	137.5%	1,631	9,359	473.8%
MA/COM	9,300	32,376	248.1%	8,680	24,305	180.0%	620	8,071	1201.8%
PHYS/AST	13,727	35,083	155.6%	12,605	27,998	122.1%	1,122	7,085	531.5%

Source: NSF, Survey of Doctorate Recipients, 1999.

Note: S&E includes engineering, the life sciences, the earth/environmental sciences, chemistry, the mathematical and computer sciences, and physics and astronomy.

foreign educated are disproportionately represented among individuals making exceptional contributions to S&E in the United States.[5] There are several reasons why the foreign born may contribute disproportionately. First, and depending on immigration law in effect at the time of entry, a work permit can require an employer declaration that the scientist is especially talented. Second, given the personal sacrifices immigration requires, immigrant scientists are likely to be highly motivated. Third, foreign-born scientists and engineers who come to the United States to receive training, especially at the doctoral or postdoctoral levels, are typically among the most able of their contemporaries. Often they have passed through several screens: they have been educated at the best institutions in their countries, withstanding intense competition for the limited number of slots available; and they have competed with the best applicants from many countries, including those from the United States, before being selected for further training in the United States (Rao 1995; Bhagwati and Rao 1996). Finally, there is some evidence that suggests that the average quality of U.S.-born individuals choosing to get doctorates in S&E declined during the 1960s, 1970s, and 1980s (Stephan and Levin 1992). This was brought about by a phenomenal growth that occurred in Ph.D. production in the 1960s and early 1970s, which arguably diluted the talent pool in science, followed by a brain drain as bright students sought more lucrative careers in business, law, and medicine.[6]

Here we present data from a study that we did in the mid-1990s to address the issue of contribution.[7] We then examine how these trends may have changed, drawing on data from the National Academy of Sciences.

We use six different indicators of exceptional work in S&E to test the hypothesis of disproportional contribution: individuals elected to the National Academy of Sciences (NAS) and/or the National Academy of Engineering (NAE), authors of citation classics, authors of "hot papers," the 250 most-cited authors, authors of highly cited patents, and scientists who have played a key role in launching biotechnology firms. We do not claim that this list is exhaustive, merely illustrative.

Members of the NAS and NAE are elected in recognition of their distinguished and continuing contributions to knowledge. We included 1,554 members of the NAS and 1,706 members of the NAE in the study.[8] Citation classics are journal articles that, according to the Institute of Scientific Information (ISI), which published them biweekly in *Current Contents*, have a "lasting effect on the whole of science."[9] We chose the 138 papers declared classics by ISI during the period June 1992 to June 1993 in the

areas of life sciences; agriculture, biology, and environmental sciences; physical, chemical, and earth sciences; and clinical medicine.[10] Authors of citation classics were considered to have made a significant contribution to
science in the United States if the author was working in the United States
at the time the article was published. In terms of author order as listed on
the publication, this resulted in the identification of 62 first authors (54
unique) and 135 non-first authors (127 unique).

Each issue of *Science Watch*, also published by ISI, contains a list of the
ten most cited or "hot papers" in chemistry and physics or medicine and
biology. The selection is based on the number of times a paper has been
cited by other authors in a given period, usually the two months prior to
the cover date of *Science Watch*. We chose the 251 papers declared "hot"
between January 1991 and April 1993. Again, an author was considered to
have made a significant contribution to U.S. science if the author was
working in the United States at the time the article was published. This
resulted in the identification of 170 first authors (161 unique) and 786 non-
first authors (686 unique).

Both citation classics and hot papers identify articles that have made or
are making a significant contribution to the knowledge base. From time to
time ISI also focuses on authors as opposed to articles, preparing lists of
the "most-cited scientists." From the list of 250 most-cited authors during
the years 1981 to 1990, we studied 183 authors who were based in the United
States.[11]

The last two criteria that we used focus on technology transfer. We studied authors of highly cited patents (the top 3.5 percent over the period
1980–91) in the field of "medical devices and diagnostics."[12] We chose medical devices because of the strong consensus that patents play a key role
in this area. We identified 206 (178 unique) U.S.-based scientists. Finally,
we identified the scientific founders and chairs of scientific advisory boards
of biotechnology firms making an initial public offering (IPO) during the
period March 1990 to November 1992.[13] We identified ninety-eight founders
and chairs (ninety-seven unique) from the prospectuses of forty firms. Altogether, the study group consisted of 4,746 scientists and engineers.

We obtained place of birth and educational origin of each scientist and
engineer, as well as the date of birth and date of degree(s), from various
scientific organizations and directories. For scientists involved with biotech
firms, we used the company's prospectus. We sought addresses for and
surveyed the 1,050 scientists and engineers for whom biographical data
could not be obtained from public sources. The response rate was 64.8

percent.[14] Overall, we ascertained essential biographical data (such as country of birth) for 89.3 percent of the study group.

We adopted an agnostic approach, despite our priors, and used a two-tail test. For each of the six indicators, we determined whether the observed frequency by birth (or educational) origin was significantly different from the frequency one would expect given the composition of the scientific labor force in the United States in either 1980 or 1990 (see Table 8.1). To do so, we used a non-parametric "goodness of fit test," computing the chi-square statistic. In cases where the chi-square statistic was inapplicable because of small cell size, a two-tailed binomial test was applied.[15] We use the 1980 benchmark for the underlying composition of the scientific workforce for individuals elected to NAS or NAE, most-cited authors, authors of citation classics, and founders/chairs of biotechnology companies, because each of these indicators was based on a list of scientific accomplishments that began before that date. The remaining indicators used a 1990 benchmark.

Engineers Elected to the NAE

Table 8.3 provides summary data concerning the birth and educational origins of engineers elected to the NAE by section. Overall, we see that the

TABLE 8.3 Birth and educational origins of members of the National Academy of Engineering (NAE) who make exceptional contributions.

Indicator	Benchmark year	Percent foreign born	Percent with foreign Bacc.	Percent foreign born of those born before 1945
All Sections *1706*	1980	19.2‡ *1705*	10.7‡ *1615*	19.3‡ *1677*
Mechanical Section *143*	1980	28.7‡ *143*	16.3‡ *135*	28.7‡ *143*
Chemical Section *141*	1980	19.9 *141*	13.4† *134*	19.9 *136*
Civil Section *217*	1980	15.7 *217*	8.3 *205*	15.7* *216*
Electrical Section *411*	1980	22.6‡ *411*	11.1‡ *386*	22.9‡ *407*
Industrial Section *85*	1980	12.9 *85*	11.1‡ *85*	13.4* *82*
Other Sections *709*	1980	17.1‡ *708*	9.3† *674*	16.9 *693*

Notes:
Bacc. = Baccalaureate degree.
Numbers of observations are in italics.
Chi-square tests of observed and expected frequencies are used. If the expected frequency is <5, and the test is inapplicable, a two-tailed binomial test is used.
* p = .05 or less.
† p = .01.
‡ p = .001 or less.

proportion of foreign-born engineers among this elite group is 19.2 percent and is significantly different from the underlying benchmark population (13.9 percent) at the P=0.01 level or less. Members of the NAE are also more likely to be educated abroad than is the underlying population (10.7 percent vs. 7.4 percent). The results and level of significance vary somewhat by field both for birth origin and educational origin, especially in the case of civil engineering, where neither proportion is significantly different from the benchmark population. The engineering section with by far the largest proportion born and educated abroad is mechanical engineering.

Scientists Making Exceptional Contributions to the Life Sciences

Table 8.4 displays data for scientists making exceptional contributions in the life sciences.[16] Included are separate indicators for first and non-first

TABLE 8.4 Scientists making exceptional contributions in the life sciences in the United States.

Indicator	Benchmark year	Percent foreign born		Percent foreign educated			
				Bacc.		Ph.D.	
Citation classics, first authors 43	1980	27.5†	40	18.4	38	15.0	40
Citation classics, not-first authors 104	1980	22.7*	75	16.2	74	14.1	71
Highly-cited patents, medical devices 178	1980	17.6‡	74	11.1†	72	n/a	n/a
Most-cited authors 164	1980	29.1‡	151	19.4‡	144	21.7‡	152
Outstanding authors 204	1980	28.7‡	188	18.5†	178	20.1‡	189
NAS members 744	1980	21.1‡	733	9.1†	646	12.4‡	712
Founders/chairs biotech co.s 97	1980	24.7†	81	16.9	77	14.1	92
Hot papers, first authors 74	1990	17.8	45	13.6	44	10.6	47
Hot papers, not-first authors 388	1990	22.6	235	16.3	221	12.4	226

Notes:
Bacc. = baccalaureate degree; Ph.D. = doctoral/medical degree; n/a = not applicable.
Numbers of observations are in italics.
Chi-square tests of observed vs. expected frequencies are used. If the expected frequency is <5, and the test in inapplicable, a two-tailed binomial test is used.
* p = .10 or less.
† p = .05 or less.
‡ p = .01 or less.

authors of citation classics and hot papers, members in NAS sections in the life sciences, and a category called "outstanding" authors, which combines the most cited with the first authors of citation classics. We find that all indicators benchmarked by the 1980 composition of the scientific labor force are statistically significant at the P=0.10 level or less, with several at the P=0.01 level or less. Nearly three out of ten of the "outstanding" authors are foreign born, compared to a population percentage of 15.4. The proportion of foreign born among first and non-first authors of hot papers is not, however, significantly different from the proportion found for life scientists in the 1990 benchmark population.

The proportion of foreign-educated life scientists making exceptional contributions is significantly different from the benchmark population at the baccalaureate and doctoral degree level in the case of most-cited and outstanding authors and for members of the NAS. The same is true at the baccalaureate level in the case of those authoring highly cited patents for medical devices.

Scientists Making Exceptional Contributions to the Physical Sciences

Regardless of benchmark data or indicator, we find the foreign born to be disproportionately represented among those making exceptional contributions in the physical sciences (Table 8.5). For example, more than half (55.6 percent) of the "outstanding" authors in the physical sciences are foreign born compared to just 20.4 percent of physical scientists in the scientific labor force as of 1980 (Table 8.1). We also find that the foreign educated are disproportionately represented for a number of the indicators—among most-cited and outstanding authors as well as first authors of hot papers.

Update of Earlier Findings

We update the study by examining the composition of the National Academy of Sciences (NAS) by birth and educational origin for the year 2004, comparing the 2004 results to the composition of the National Academy in 1994 (Table 8.6). For purposes of analysis, we exclude emeritus and honorary members and foreign associates lacking a U.S. institutional address. The table indicates that while the NAS population we include in the analysis has grown slightly in the ten-year interval, the number of foreign-born members has remained almost constant.[17] The percentage of foreign born has thus declined. We also find that the number receiving their doctoral training abroad has declined in absolute terms, as has the percentage.[18] The same pattern holds regarding birth origin and origin of doctoral training

TABLE 8.5 Scientists making exceptional contributions in the physical sciences in the United States.

Indicator	Benchmark year	Percent foreign born		Percent foreign educated			
				Bacc.		Ph.D.	
Citation classics, first authors *11*	1980	§	§	§	§	§	§
Citation classics, first and not-first authors *34*	1980	40.9†	*22*	21.1	*19*	33.3‡	*21*
Most-cited authors *19*	1980	64.7‡	*17*	56.3‡	*16*	31.3‡	*16*
Outstanding authors *29*	1980	55.6‡	*27*	41.7‡	*24*	30.8‡	*26*
NAS members *474*	1980	26.7‡	*465*	13.0	*429*	11.4‡	*458*
Hot papers, first authors *87*	1990	35.5†	*76*	28.4†	*74*	18.1*	*72*
Hot papers, not-first authors *299*	1990	35.4‡	*192*	23.4*	*188*	13.0	*177*

Notes:
Bacc. = Baccalaureate degree; Ph.D. = doctoral/medical degree.
Numbers of observations are in italics.
Chi-square tests of observed vs. expected frequencies are used. If the expected frequency is <5, and the test is inapplicable, a two-tailed binomial test is used.
* p = .10 or less.
† p = .05 or less.
‡ p = .01 or less.
§ Combined with non-first authors because of sample size.

within the five broad NAS sections with some exceptions: in the math/ computer science sections, the number of foreign-born members has grown in absolute terms, but the overall growth of members dwarfs this effect, leading to a decrease in the percent foreign born; in engineering, the total number has declined, but the number of foreign born has declined even more; and in earth sciences the absolute number has declined, as has the number of foreign born.

The absolute and percent declines need not translate into a rejection of the hypothesis that exceptional contributions are disproportionately drawn from the foreign born and foreign educated. That depends on the underlying distribution of the foreign born and foreign educated. Chi-square tests indicate the foreign born to be disproportionately represented in all NAS sections in which they were disproportionately represented in 1994. The findings are somewhat different in terms of educational origin. In the earth/environmental sciences section, we now find the native educated to be disproportionately represented, while only members of the physical

TABLE 8.6 Birth and educational origins of members of the National Academy of Sciences (NAS) by section, 1994 and 2004.

Year	NAS section	Foreign born			Foreign Ph.D.		
		Number		Percent	Number		Percent
1994	All sections (1554)	363	*1524*	23.8‡	168	*1467*	11.5‡
	ENG (84)	21	*83*	25.3‡	2	*75*	2.7
	LIFE SCI (744)	155	*733*	21.1‡	88	*712*	12.4‡
	PHYS SCI (474)	124	*465*	26.7‡	52	*458*	11.4‡
	MATH/COMP (128)	40	*123*	32.5‡	11	*107*	10.3
	EARTH/ENVIRON (124)	23	*120*	19.2	15	*115*	13.0
2004	All sections (1716)	359	*1716*	20.9‡	151	*1700*	8.9
	ENG (75)	17	*75*	22.7†	5	*74*	6.8
	LIFE SCI (814)	146	*814*	17.9†	68	*806*	8.4
	PHYS SCI (536)	129	*536*	24.1†	56	*533*	10.5‡
	MATH/COMP (174)	51	*174*	29.3‡	17	*170*	10.0
	EARTH/ENVIRON (117)	16	*117*	13.7	5	*117*	4.3^

Notes:
Group size is in parentheses; numbers of observations are in italics.
Chi-square tests of observed and expected frequencies are used. If the expected frequency is <5, and the test is inapplicable, a two-tailed binomial test is used.
* p = .10 or less.
† p = .05 or less.
‡ p = .01 or less.

ENG	engineering	MATH/COMP	math and computer sciences
LIFE SCI	life sciences	EARTH/ENVIRON	earth and environmental sciences
PHYS SCI	physical sciences		

sciences sections are disproportionately foreign educated in 2004. By way of comparison, in 1994 we found members of the life sciences, physical sciences, and "all" NAS members to be disproportionately foreign educated in terms of doctoral education.

We choose the 1980 benchmark for the chi-square tests, given that election is based on previous work and the mean and median years of election for the sections fall before 1990.[19] If we had, instead, chosen the 1990 benchmark, we would have come up with substantially different conclusions. To wit, we would have found the native born to be disproportionately represented among all members as well as members of the life science sections and disproportionately represented (although at a lower level of significance) among members of the engineering section. In the other four fields, one cannot reject the hypothesis that the distributions are the same as the underlying population.

Our findings suggest that the United States may be in a transition period in terms of the contribution of the foreign born. It is difficult to know why. One possibility relates to a change in the underlying age distribution. The dramatic increase in the proportion of foreign born working in the United States is due in large part to the immigration of young scientists and engineers. These younger foreign-born scientists and engineers have a lower probability of having made exceptional contributions or, if they have, of being recognized for their contributions by 2004. We do not know if our findings would change if we were to focus on different indicators of exceptional contribution, especially indicators that have a higher representation of "young" scientists and engineers than does the NAS. We do note, however, that in our earlier work (Stephan and Levin 2001) the proportion of foreign born among first and non-first authors of hot papers in the life sciences was not significantly different from the proportion found for life scientists in the 1990 benchmark population. That proportion was, however, significantly different for the physical scientists. These findings are especially relevant since the hot paper indicator picks up significant work soon after it is published; thus the young need not wait until middle age to be recognized as having made a significant contribution. When we make the tests for hot papers conditional on age, the results persist for the younger group. The benchmarks were not, however, stratified across multiple age categories, and the results could change if we had more categories.[20]

The apparent transition may also be an artifact of events that transpired more than sixty years ago. As noted above, the number of foreign born making exceptional contributions, as measured by membership in the National Academy of Sciences, experienced at most modest growth during the period 1994–2004. This slowed growth may be the result of the extremely large number of foreign-born exceptional contributors, created in part by events in Europe in the 1930s, who worked in the United States during the second half of the twentieth century. Many of these exceptional contributors are now either retired or deceased. In a related way, the results may reflect changes in the cohort composition of the NAS during the ten-year interval. In our earlier work we found noncitizens who were born prior to 1945 to be disproportionately represented among members of the 1994 NAS. We did not find a comparable result for those born in 1945 or later.

We also note several trends afoot that could influence the number of highly productive foreign-born scientists working in the United States. Four are of particular note. First, certain countries have created programs designed to reverse the expatriate brain drain. China is a case in point, but such policies also exist in many European countries as well as in Taiwan and Australia.[21] Recently there has also been considerable fanfare regarding the large number of Indian computer scientists returning to India to establish companies or work for newly created companies. While the Indian government has not necessarily recruited computer scientists to return, attractive opportunities in India coupled with the dot-com crash and tightened U.S. immigration rules make returning an appealing alternative. Second, mandatory retirement policies are leading some eminent older European scientists to work in the United States. A case in point is the Swiss chemist Kurt Würthrich. In anticipation of retirement, Würthrich had already switched much of his research operation to the Scripps Research Institute in La Jolla, California, prior to winning the Nobel Prize in 2002. There are other examples of talented older researchers who choose (and are invited) to come to the United States rather than be forced to retire (Weiss 2003). It is not only that these scientists can continue to work; they can also be principal investigators (PIs) and garner resources for research. The lack of resources for research constitutes a third factor affecting the balance of foreign-born talent working in the United States (Carvajal 2004). This lack of resources has been particularly felt in France and Germany in recent years (Carvajal 2004). Finally, the post-9/11 changes in immigration policy may be affecting the number of foreign-born scientists studying in and subsequently working in the United States.

Discussion of Contribution

Our results indicate that although there is slight variation by discipline, individuals making exceptional contributions to U.S. S&E in the recent past were disproportionately drawn from the foreign born. Only in the instance of hot papers were we unable to reject the null hypothesis that the proportion was the same as that in the underlying population and then only for the life sciences. We also find evidence that, for a number of criteria, individuals making exceptional contributions to U.S. S&E in the recent past were disproportionately drawn from the foreign educated, both at the undergraduate and graduate level. The 2004 NAS results suggest that a transition may be in process and that these conclusions concerning contributions of the foreign born may not hold in the future.

We conclude that the United States has benefited from the inflow of foreign-born talent and that this talent was more likely to have been educated abroad than one would have predicted given the incidence of foreign-educated scientists in the scientific workforce. Thus, to the extent that contributions in S&E are geographically bounded, as a country the United States has benefited from the educational investments made by others.

It does not necessarily follow, however, that these benefits have been produced at no cost or at a low cost to U.S. citizens. We investigate the issue of incidence in the next section of this study, focusing on how the increased supply of foreign-born scientists and engineers affects the labor market outcomes of native-born scientists and engineers.

THE ISSUE OF INCIDENCE

Although there is a "widespread perception that 'immigrant hordes' have an adverse effect on the employment opportunities of U.S. citizens" (Borjas 1994, 1667), the question of how immigrants affect employment outcomes in S&E has yet to be investigated.[22] To date, the evidence is sketchy, consisting of anecdotal reports and selected data, implying that in some fields immigrants "take" coveted positions away from U.S. citizens in science, especially in academe. For example, the American Mathematical Society noted that "immigrants won 40% of the 720 mathematics jobs available last year (1995) . . . and helped boost the unemployment rate into double digits among newly minted math Ph.D.s" (Phillips 1996, A2). And a study by the National Research Council reported a growing "imbalance between the number of life-science Ph.D.s being produced and the availability of positions that permit them to become independent investigators," a situation exacerbated by the "influx of foreign-citizen Ph.D. candidates" (1998, 4).[23]

Here we analyze the differential employment patterns of U.S.-doctorate recipients in S&E over the period 1979–97[24] using data from the SDR. We seek to determine how U.S.-citizen S&E doctorates have fared relative to their noncitizen counterparts and, in particular, whether there is evidence of substitution of noncitizens for citizens, especially in the academic sector. Although the SDR excludes two groups that are important to the scientific workforce, scientists and engineers working in the United States who received their doctoral training abroad and scientists with medical degrees who lack U.S.-earned doctoral degrees, it remains the best available data source for the purpose of studying changing patterns over time.

Methodology

To tackle the question of incidence, we undertake a thought experiment. We compare the *actual* employment growth of a specific "citizenship" group (citizen or noncitizen) in a specific sector with the amount *predicted* using the following counterfactual. We ask what would have happened to employment of U.S.-citizen (noncitizen) S&E doctorates in different sectors of the economy if their employment had grown at the overall growth rate for all S&E doctorates combined, regardless of citizenship status. In doing so, we acknowledge that the growth in U.S.-trained S&E doctorates has been fostered both directly and indirectly by a variety of policies, including changes in immigration laws and the widespread availability of funds supporting graduate and postdoctoral study in science. In effect, we assume that the United States could have implemented a different set of policies that would have elicited an equal amount of growth from citizens alone. Whether this is the "correct" counterfactual is, of course, subject to debate. But the belief exists that "the United States should be able, if it so chose as a matter of social policy, to meet its needs for scientists from within its own population, especially by harnessing the talents of under-represented minorities and women" (Bouvier and Martin 1995, 3).[25]

We implement the analysis by adapting a technique originally developed in the regional science literature known as shift-share.[26] The conventional (regional science) application of shift-share decomposes employment growth for industry i in region j, G_{ij}, into three components: (1) a reference group or "overall" growth component (such as employment growth in the United States), O_{ij}; (2) an industrial-mix component, M_{ij}; and (3) a "competitive" component, C_{ij}. In the present analysis, the reference group is U.S.-S&E doctoral recipients; "regions" refers to the employment sectors of S&E doctorates (academe, nonacademe, and other); and "industries" refers to the citizenship status of S&E doctorates (citizen or noncitizen).

For each citizenship group in each sector, the following identity must hold:

$$G_{ij} - O_{ij} = M_{ij} + C_{ij}$$

where

$$O_{ij} = b_{ij} \, r_{oo}$$
$$M_{ij} = b_{ij} \, (r_{io} - r_{oo})$$
$$C_{ij} = b_{ij} \, (r_{ij} - r_{io})$$

and b_{ij} = employment for citizenship group i in sector j during the base period, r_{oo} = the overall growth rate for all S&E doctorates, r_{io} = the growth rate for citizenship group i, and r_{ij} = the growth rate for citizenship group i in sector j.

Thus $G_{ij} - O_{ij}$ measures the difference between the *actual* growth in employment and the *predicted* growth in employment for group i in sector j; the difference is then divided into M_{ij}, now termed the "minting" effect, and C_{ij}, the competitive effect. The minting effect measures the employment change citizens (noncitizens) experienced in a particular sector due to the differential in growth rates between its doctoral recipients and all doctoral recipients. By definition, the minting effect must sum to zero for the two citizenship groups. The competitive effect is the difference between the actual change in employment for each citizenship group in each sector and the employment growth that would have occurred had each group grown at its overall growth rate. By analogy, as in the case of international trade, competitive effects across sectors for a particular group (citizen or noncitizen) must sum to zero just as trade accounts must balance out. In addition, subsector additivity must hold. That is, for each citizenship group, if a sector such as academe is partitioned into two or more parts, the sum of the competitive effects for all parts must equal the competitive effect for the sector as a whole.

In effect, C_{ij} captures the differential rate at which jobs in various sectors of the economy have grown for each citizenship group, after accounting for the overall growth in the number of doctoral recipients and the differential minting effects observed. We define the substitution effect to be the difference between the citizen and noncitizen competitive effect. Thus, suppose we observe that employment growth for citizens in academe is smaller than predicted given the counterfactual. There are two reasons why this may have happened: the citizen share of S&E doctorates may have declined (the minting effect), or citizens may have experienced slower employment growth in academe than in the other sectors (the competitive effect). To determine how citizens compare to their noncitizen counterparts in academe—whether substitution has occurred—we then subtract the noncitizen competitive effect from the citizen competitive effect (both measured in percentage terms to adjust for relative size differences).

Although the decomposition into a minting effect and a competitive effect is based on an accounting identity, from a public policy standpoint these are powerful distinctions to make since the prescriptions for remedy differ. For example, to the extent that the minting effect works against citizens, efforts are needed to help expand their numbers in doctoral programs. To the extent that the competitive effect works against citizens relative to their immigrant counterparts, then policy makers need to consider whether their movement from academe is of an involuntary or voluntary nature. Have U.S. citizens been pushed out of positions in academe by the inflow of foreign talent, or have they been pulled out by the lure of better salaries and opportunities in other sectors?

Substitution from Academe

Table 8.7 presents substitution estimates in the academic sector obtained from the decompositions performed for all fields combined and major subfield over the period 1979–97.[27] This sector includes individuals under the age of sixty-five who are either employed full time or hold a postdoctoral training position in a university, four-year college, or medical school. The negative competitive effects for both citizens and noncitizens indicate that both groups have lost employment share in academe relative to the remaining sectors in the analysis—nonacademe and other. Moreover, for each field, and without exception, using the above definition, we find that noncitizens have been substituted for citizens in academe, since the citizen (negative) competitive effect is larger in absolute value than the (negative) competitive effect for the noncitizen. This effect is largest for citizens in the life and physical sciences. Within the physical sciences, substitution is largest for those in the mathematical and computer sciences.

Substitution within Academe

Not only has employment in academe fallen relative to the other two sectors for citizens and noncitizens alike, the types of appointment held by both citizenship groups have changed as well. Figure 8.2 examines substitution *within* academe where the type of appointment is partitioned into "faculty" vs. "postdocs." Figure 8.3, however, examines substitution *within* academe where the type of appointment is partitioned into "permanent"— tenured or tenure-track faculty—versus "temporary"—postdocs and other non–tenure track and nonfaculty positions such as lecturers, instructors,

TABLE 8.7 Substitution from academe, 1979–97.

| | Competitive effects | | |
Field	U.S. citizens	Non-citizens	Substitution*
All fields combined	−13.9%	−6.8%	−7.1%
Engineering	−16.3%	−8.8%	−7.5%
Life sciences	−11.4%	−0.7%	−10.7%
Physical sciences	−19.6%	−8.2%	−11.4%

* Calculated as the competitive effect (%) for U.S. citizens less the competitive effect (%) for non-citizens.

clinical faculty, research scientists, and technical staff. Again, we restrict the analysis to those who are full time and under the age of sixty-five.

Figure 8.2 shows that overall, for all fields combined and in the life sciences, the substitution of noncitizens for citizens in academe can primarily be attributed to their substitution into postdoctoral appointments and not into faculty positions. Indeed, there is minimal evidence of noncitizens being substituted for citizens in faculty positions (-1.7 percent) for all fields taken together, and in the life sciences citizens have actually fared relatively better than noncitizens (+5.3 percent) have when considering faculty appointments. This is not true, however, in engineering and in the physical sciences. Here we find that the substitution of noncitizens for citizens in academe is largely accounted for by their substitution into faculty positions and not into postdoctoral positions.

But, as Figure 8.3 illustrates, the story is somewhat different when one considers who holds permanent versus temporary appointments within the academic sector. Now we see that for all fields taken together as well as for each subfield, the substitution effect in academe observed for noncitizens can primarily be attributed to their substitution into temporary rather than permanent positions. Moreover, for all fields taken together, there is scant

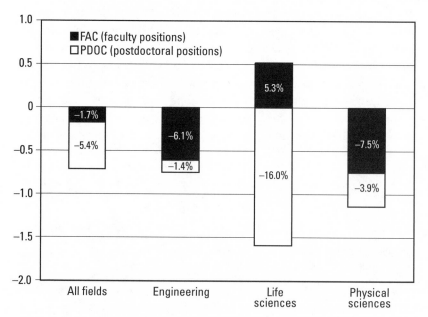

FIGURE 8.2 Substitution within academe, faculty vs. postdocs, 1979–97.

evidence of substitution of noncitizens for citizens in permanent academic appointments (-0.6 percent), and in the life sciences citizens have again fared relatively better than noncitizens (+1.6 percent) in terms of holding permanent academic appointments.

Discussion of Substitution Effects

Our analysis indicates that both citizens and noncitizens experienced employment shortfalls in academe (negative competitive effects) after accounting for the overall growth in the number of doctoral recipients and the differential rate at which the two groups minted degrees. Citizens, however, fared relatively worse than their noncitizen counterparts, and, by our definition, substitution has occurred. But citizen S&E doctorates, except in the physical sciences and engineering, have been more successful than noncitizens in holding faculty positions rather than postdoc positions within academe. Furthermore, citizen S&E doctorates have generally been more successful than their noncitizen counterparts in holding permanent, tenured, or tenure-track faculty positions, rather than positions as temporary members of the academic units.

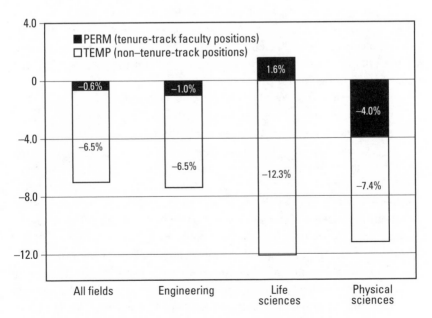

FIGURE 8.3 Substitution within academe, permanent vs. temporary, 1979–97.

Our analysis cannot determine whether substitution occurred on balance because citizens were pushed out by the heavy inflow of foreign talent or pulled out by the better opportunities that were available to them elsewhere in the economy. The finding that the substitution effects observed in academe can be largely attributed to the substitution of noncitizens for citizens in postdoc and other temporary appointments suggests an element of pull. Specifically, noncitizens may have been substituted for citizens in these less desirable positions because they are less able than citizens to respond to the lure of better opportunities elsewhere. For example, citizens do not face the visa constraints that can be encountered by noncitizens, affecting their transition from training positions to more permanent positions.

Conclusion

We conclude that the foreign born, at least in the recent past, have contributed disproportionately to U.S. science. Moreover, a surprisingly large percentage of the foreign born working in the United States were educated abroad, suggesting that the United States has benefited from investments made by other countries. Our evidence, however, is somewhat dated, resting on indicators of contribution collected in the early 1990s. The 2004 NAS results suggest that a transition may be in process and that the conclusions concerning the foreign born may not hold in the future. Indeed, if the 1990 benchmark were to be used, in several instances we find the native born to be disproportionately represented.

Our work suggests that the benefits have not been without costs. One group that may have borne the costs is citizen-scientists and citizen-engineers, who have fared relatively worse than their noncitizen counterparts in terms of holding jobs in academe. The costs borne by native-born scientists and engineers, however, are mitigated in two ways. First, we have seen that substitution occurs mostly in "temporary" jobs in academe, not in "permanent" jobs in academe. Thus, noncitizen-scientists are being substituted for citizen-scientists in the less valued as opposed to more highly valued positions within the academic community. Second, this result together with the finding that substitution is largest for those in the mathematical and computer sciences suggests that citizen-S&E doctorates, at least in certain fields, have been pulled out by higher paying jobs and not pushed from the academic sector. In other words, citizen-scientists appear to leave the academic sector to seek better opportunities and higher paying positions elsewhere in the economy. To the extent this is the case, noncitizens bear the costs brought about by an increased supply in the science and engineering workforce by working in jobs that native-born scientists and engineers eschew because of issues related to salary and job security.

For the United States to remain competitive, it is important that the country attract exceptionally talented individuals to work in science and engineering (National Academy of Science 2005). Some of these talented individuals are citizens; others are noncitizens. A different set of relative rewards and opportunities could alter the proportion of citizens attracted into science and engineering. Noncitizens are also responsive to the rewards and opportunities, as is evidenced by the large increase in recent years of temporary residents in areas with substantial funding, such as the life sciences. But noncitizen participation in science and engineering is also affected by governmental policies. As noted in chapter 6, and in a 2005 National Academy of Science Report, there is some evidence that the number of noncitizens coming to the United States to study science and engineering declined in the wake of 9/11.[28] A substantial decline in the number of noncitizens coming to the U.S. could dramatically affect the productivity of the U.S. science and engineering enterprise.

Notes

1. In this analysis, citizens include those naturalized or native born at the time the doctorate was earned; noncitizens include permanent and temporary residents and individuals who had applied for citizenship at the time the doctorate was earned.

2. The NSCG was fielded in 1993 and collected information on the education and labor market experiences of college-educated individuals identified in the 1990 U.S. Decennial Census. Table 8.1 is drawn from Stephan and Levin (2001).

3. We could have used the 1982 Postcensal Survey (1982 Survey of Natural and Social Scientists and Engineers) for the 1980 estimates. We chose not to do so, however, because the NSCG was a superior survey, having supplemented the mail-only questionnaire with telephone interviews and intensive follow-ups to nonrespondents.

4. In recent years, many of the foreign doctorate recipients working in the United States initially came to take a postdoctoral position. It should also be noted that some of those who have foreign doctoral degrees are U.S. citizens who go abroad for training. This is most common in the earth and environmental sciences.

5. Portions of this section draw heavily on Stephan and Levin (2001).

6. There is evidence that the decline in quality continued through the 1990s. A

study sponsored by the Sloan Foundation (Best and Brightest 2000) found that among U.S.-citizen GRE test-takers scoring 700 or above on the quantitative test, the number going on to graduate school in an S&E field had declined during the periods 1987–88 and 1997–98.

7. For a detailed discussion, see Stephan and Levin (2001).

8. From the 2,075 NAS members in 1994, we excluded foreign associates without a U.S. address, Public Welfare Medalists (who are honorary members), members of the psychology and social science sections, and twenty members for whom no section was specified. From the 1,781 members of the NAE as of June 30, 1995, we excluded foreign associates without a U.S. address.

9. ISI discontinued the practice of declaring Citation Classics in the late 1990s.

10. We excluded papers published before 1970 because of the difficulty in obtaining biographical information for authors.

11. David Pendlebury of ISI provided the list. Heavily cited authors with common last names were omitted from the list because ISI could not accurately determine attribution.

12. The list was prepared by Francis Narin of CHI Research, Inc., using the database created by the company. Research suggests that citations to patents (the citations that appear on the front page of a patent under "references cited") can be used as an index of the importance of a given patent. See, for example, Albert et al. (1991).

13. Individuals were assumed to be scientists if they held either the Ph.D. or M.D. degree. Audretsch and Stephan (1996) examine the various roles that scientists play with start-up firms in biotechnology.

14. The response rate was 54.5 percent for the entire sample and 64.8 percent for deliverable surveys. A review of the names of the nondeliverables suggests that a disproportionate number may have been foreign born. For the nonrespondents there does not appear to be a birth origin bias.

15. The chi-square statistic is inapplicable when the expected frequency in any cell is less than 5 and there are just two categories in the classification of the data. In such cases, the binomial test is uniquely applicable. See Siegel (1956, 59).

16. We include biology and medicine in the life sciences. The physical sciences include chemistry and physics.

17. Note that the size of group and information n for place of birth are the same for 2004; for 1994 there were some members of the NAS for whom country of birth could not be determined, and hence the information n is smaller than the size of group for place of birth.

18. The NAS data on education (degrees, years, institutions) is subject to measurement error. We have made every effort to verify the educational data when possible.

19. The 1980 benchmark is used for comparisons with the 1994 NAS as well. Although 1970 benchmarks would have arguably been more appropriate, they were not available to the researchers.

20. The test divided the groups into those who were under forty-eight and those who were forty-eight or over at the time the papers were declared hot.

21. China has initiated several programs (Hundred People, 300 Talents, and Changjiang Scholars) especially designed to lure Chinese scientific talent back to China (Normile 2000). See Carvajal (2004) for a discussion of what other countries have done.

22. This section draws on work by Levin, Black, Winkler, and Stephan (2004a and 2004b).

23. U.S.-citizen information technology (IT) workers also claim that the increased flow of H1-B visa holders are adversely affecting their careers in IT (Matloff 1988).

24. While data are available from 1973, we start with 1979 because of the poor quality of the survey questions concerning tenure status and academic rank found in the 1973 SDR (Levin and Stephan 1991).

25. North (1995) observes that "while the large-scale presence of foreign-born S/Es, particularly at the Ph.D. level, was neither deliberately created by America's universities and corporations nor thrust upon them against their will (145) . . . their presence and growing numbers are . . . permitting the status quo to continue without the awkward adjustments that would be needed were they not here" (161).

26. See, for example, Gordon, Hackett, and Mulkey (1980); Andrikopoulos, Brox, and Carvalho (1990); Kiel (1992); Grobar 1996. In recent years, shift-share has been applied in a wide variety of contexts, including Smith (1991), Ishikawa (1992), Geiger and Feller (1995), Haynes and Dinc (1997), and Hoppes (1997).

27. To conduct the analysis, the data are initially divided into three six-year intervals (1979–85, 1985–91, 1991–97). Then each component in the decomposition for each time period is summed over the three periods so that a single number captures the "dynamic" nature of employment growth for the entire 1979–97 period. See Barff and Knight (1988) for insight into this procedure. Because beginning in 1991 several changes were made to the SDR in an attempt to increase its response rate, we use the older, mail-only weight for the interval 1985–91 for better comparability with the pre-1991 data. In the shift-share analysis the definition of the physical sciences is broadened to include math and computer science and earth/environmental sciences as well as physics and chemistry.

28. Other factors contributed to the decline, such as increased competition from programs outside the United States and a decline in employment opportunities following the recession in the early 2000s.

U.S.-Citizen Ph.D.s in Science and Engineering

9

The Changing Composition of U.S.-Citizen Ph.D.s

JEFFREY A. GROEN AND MICHAEL J. RIZZO

INTRODUCTION

American research universities are admired at home and abroad. As Grant Black and Paula Stephan (chapter 6 herein) indicate, doctoral programs at U.S. universities increasingly attract students from other countries, especially in science fields. As a consequence, the share of doctorates at U.S. universities awarded to U.S. citizens has fallen substantially over the previous four decades. In this chapter, we describe changes in the composition of U.S.-citizen doctorate recipients from the early 1960s to 2000. We examine composition in terms of fields of doctoral study, type of undergraduate institution, gender, and race and ethnicity.

Doctorate recipients represent the primary source of new talent in many occupations and professions. Doctorates in science and engineering are of particular interest to researchers and federal policymakers because the size and quality of the nation's scientific workforce are perceived to affect the pace of innovation and economic growth. Of particular interest for science and technology policy is how the number and composition of new Ph.D.s responds to the wages of scientists and engineers.

The flow of new doctorates is also relevant for our nation's colleges and universities in several ways. First, these institutions train future Ph.D. recipients at the undergraduate level. Describing the undergraduate origins of doctorate recipients illustrates the role of different types of institutions in preparing and inspiring their students to enter Ph.D. programs. Second, the demographic composition of new Ph.D. recipients reflects in part the ability of graduate programs to attract increasing numbers of women and racial minorities. Third, since colleges and universities themselves employ a large share of new Ph.D. recipients as professors, the composition of doctorates is relevant for understanding the composition and structure of academic labor markets.

The next section of this chapter describes overall graduate education trends among U.S. citizens, including the number of doctorates and the proportion of those with bachelor's degrees going on to earn Ph.D.s. The chapter then turns to changes in composition along various dimensions: field of study, type of undergraduate institution, gender, and race and ethnicity. Throughout the chapter, we pay particular attention to trends within science fields. Since we cover considerable ground, our approach is primarily descriptive. However, we also point toward potential explanations for some of the trends. While we are not the first to discuss the backgrounds of doctorate recipients, we do present a rather complete picture of changes along many dimensions of doctoral flows to U.S. citizens over the past forty years.[1]

Data and Overall Trends for U.S. Citizens

Data on Doctorate Recipients

Our data on the number and characteristics of individuals receiving doctorates are based on the Survey of Earned Doctorates, the data set used by Black and Stephan in their chapter and by Freeman, Jin, and Shen in theirs. The survey, conducted since 1958, is an ongoing census of all individuals earning research doctorates at U.S. universities.[2] The surveys are completed by doctorate recipients once they have satisfied the requirements for their degrees. The survey collects information on demographics, including gender, citizenship, ethnicity, and racial group; education history, including field of degrees; sources of graduate student support; employment status during the year preceding receipt of the doctorate; postgraduation plans; and parents' education. Since almost all doctorate recipients complete and return the survey, it is the most reliable source of national data on doctorate recipients.[3] Here we examine trends in doctorate recipients since 1963.[4]

Trends by Citizenship

As a context for analyzing trends among U.S. citizens, we first present trends in the total number of doctorates conferred, including those awarded to U.S. citizens and non–U.S. citizens. The total number of Ph.D.s conferred by U.S. universities has increased substantially over the past forty years, rising from 12,720 in 1963 to 41,368 in 2000 (Figure 9.1).[5] The time trend can be broken into three parts. From 1963 to 1973, the number of Ph.D.s conferred grew rapidly (at an average rate of 10 percent per year) and the growth was fueled by Ph.D.s awarded to U.S. citizens. Then the number of Ph.D.s conferred declined slightly over the next period, 1973–85. Since 1985 the number of Ph.D.s conferred has grown gradually among both U.S. citizens and non–U.S. citizens.

The increasing presence of non–U.S. citizens in U.S. doctoral education is evident in these data. The share of Ph.D.s awarded to U.S. citizens has fallen from 87 percent in 1963 to 71 percent in 2000 (Figure 9.2). The decline was concentrated over the period from 1980 to the mid-1990s. As Black and Stephan document in their chapter, the drop was particularly strong in science fields, where the U.S.-citizen share fell from 75 percent in the mid-1960s to 52 percent in 2000. (Here we distinguish between "science" and "nonscience" fields; "science" fields are defined as life sciences, physical sciences, and engineering. See the appendix to this chapter for details.) In contrast to the pattern over the previous thirty years, it is worth noting that the U.S.-citizen share increased slightly in the late 1990s. This recent trend appears to be driven by the life sciences, for which the U.S.-citizen share increased from 61 percent in 1996 to 70 percent in 2001. For the remainder of the chapter we limit our analysis to U.S. citizens.

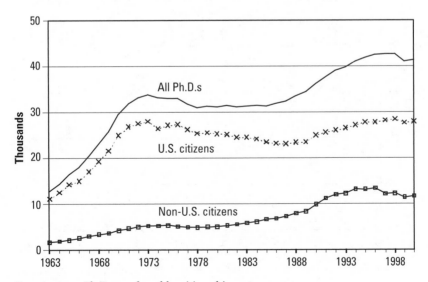

FIGURE 9.1 Ph.D.s conferred by citizenship, 1963–2000.

Source: National Science Foundation, Survey of Earned Doctorates.

Ph.D.s and Bachelor's Degrees

Given the general expansion of higher education in the United States since World War II, a useful way of understanding trends in the number of Ph.D.s conferred to U.S. citizens is by comparison to the number of bachelor's degrees awarded. The number of Ph.D.s awarded in year t, G_t can be decomposed into two terms using the following identity (Bowen, Turner, and Witte 1992): $G_t = p \times B_{t-9}$ where B_{t-9} is the number of bachelor's degrees conferred nine years earlier. Since having a bachelor's degree is typically a prerequisite for entering a Ph.D. program, B_{t-9} measures the pool of prospective Ph.D. students and is therefore a natural reference group for Ph.D. recipients in year t. The second part of the identity, p, is the ratio of G_t and B_{t-9}: the number of Ph.D.s conferred in year t as a proportion of the number of bachelor's degrees conferred nine years earlier. We call p "Ph.D. propensity" because it is a measure of the propensity for a given cohort of bachelor's degree recipients to earn Ph.D.s.

This identity is useful for interpreting trends in Ph.D.s conferred (overall and within subgroups) because it separates changes in the number of Ph.D.s conferred into those due to changes in the pool of prospective Ph.D. students (the size of the bachelor's degree cohort) and to those due to

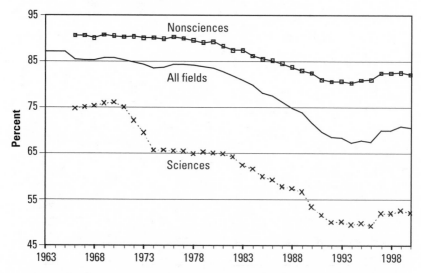

FIGURE 9.2　Ph.D.s conferred to U.S. citizens as share of total, 1963–2000.

Source: NSF, Survey of Earned Doctorates.

Note: For examples of "sciences" and "nonsciences" see Appendix A.

changes in the propensity of successive cohorts to pursue and complete Ph.D.s. In particular, patterns in Ph.D. propensity should reflect changes in the incentives to pursue graduate study, such as funding available for graduate study and job prospects in academic labor markets. Our data on bachelor's degrees awarded are taken from the Earned Degrees Conferred survey, an annual survey of colleges and universities published by the U.S. Department of Education.[6] We use a nine-year lag based on the typical number of years between bachelor's degree receipt and Ph.D. receipt. The median total time span between the baccalaureate and doctorate has risen from 8.0 years for 1966 doctoral recipients to 10.3 years for 2000 doctoral recipients. Given that students who earn their bachelor's degree in a given year take different amounts of time to earn a Ph.D., our matching of bachelor's and Ph.D. cohorts is not exact. However, it does provide a useful metric for understanding large changes.[7]

The pool of prospective Ph.D. recipients, as indicated by the number of bachelor's degrees conferred nine years earlier, has increased substantially during the postwar period (Figure 9.3). However, the increase from 265,000 bachelor's degrees in 1954 (nine years before the 1963 Ph.D. cohort) to 1.1 million bachelor's degrees in 1991 (2000 Ph.D. cohort) has not been uniform. The size of the prospective-Ph.D. pool grew steadily over the first

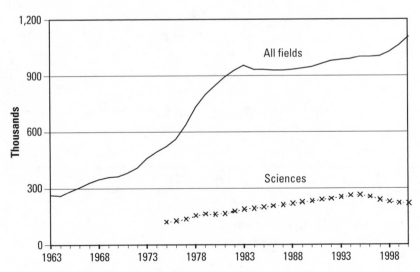

FIGURE 9.3 Bachelor's degrees conferred nine years earlier, 1963–2000.

Source: National Center for Education Statistics, *Earned Degrees Conferred.*

part of the period (1963–73 Ph.D. cohorts), increased rapidly through 1985, and increased only slightly through recent Ph.D. cohorts. Dividing the number of Ph.D.s conferred by the number of bachelor's degrees conferred nine years earlier produces an interesting pattern (Figure 9.4). The number of Ph.D.s conferred to U.S. citizens in 1963 represented 4.2 percent of bachelor's degrees conferred in 1954. Throughout the 1960s, Ph.D. propensity increased rapidly, rising to 7 percent for the 1971 Ph.D. cohort. However, Ph.D. propensity plummeted over the next decade, falling to 2.6 percent for the 1982 Ph.D. cohort. Since then it has been relatively stable at 2.5–2.8 percent.

Ph.D. propensity is greater among students in science fields. In 2000, Ph.D.s in science represented 5 percent of bachelor's degrees in science nine years earlier, while Ph.D. propensity overall was 2.5 percent. Despite this difference, since 1975 the trend in Ph.D. propensity in the sciences is quite similar to the overall trend (Figure 9.4). Throughout the period, Ph.D.

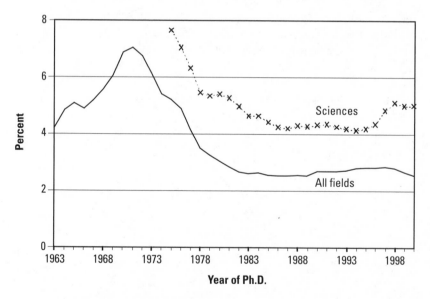

FIGURE 9.4 Ph.D. propensity, 1963–2000.

Source: Ph.D. data from NSF, Survey of Earned Doctorates. Baccalaureate data from NCES, *Earned Degrees Conferred.*

Note: Ph.D. propensity is the number of Ph.D.s conferred in a year divided by the number of bachelor's degrees conferred nine years earlier. This definition applies for all other figures in this chapter unless otherwise indicated.

propensity is roughly 2.5 percent higher among students in science. The late 1990s represent a notable exception to the common trends, however. While Ph.D. propensity was stable overall, it increased among students in science. The increase from 4.2 percent in 1995 to 5 percent in 1998 reflected both a decrease in the number of science bachelor's degrees (see Figure 9.3) and an increase in the number of science Ph.D.s.

The trends in Ph.D. propensity and bachelor's degrees conferred over all fields can shed some light on trends in Ph.D.s awarded to U.S. citizens (see Figure 9.1). For each of the three measures, there are three distinct periods of change, as noted above, and these periods roughly coincide across the measures. Over the first period, 1963 to 1971, the rapid increase in Ph.D. propensity and steady growth in the number of bachelor's degrees conferred fueled rapid growth in the number of Ph.D.s conferred. From 1971 to 1983, the number of Ph.D.s declined because of the steep decline in Ph.D. propensity, in spite of the strong growth in the number of bachelor's degrees. Since 1983 the number of Ph.D.s has increased slightly due to modest growth in the number of bachelor's degrees while Ph.D. propensity has been stable.

FIELD OF STUDY

In turning from trends in the overall number of Ph.D.s conferred to U.S. citizens to trends in the composition of new doctorate recipients along a variety of dimensions, we first consider changes in the composition of new Ph.D.s across broad fields of study. In 2000 approximately 40 percent of Ph.D.s awarded to U.S. citizens were in the three science fields: 20 percent in life sciences, 12 percent in physical sciences, and 8 percent in engineering (Table 9.1). Among the nonscience fields, approximately 20 percent of Ph.D.s were awarded in both social sciences and education, and 11 percent were awarded in humanities.

Given that the total number of Ph.D.s awarded to U.S. citizens across all fields has nearly doubled since 1966, it is perhaps not surprising that the number of Ph.D.s awarded in each of the six broad fields has increased over the 1966–2000 period. However, some fields grew faster than others. Among the science fields, life sciences grew the fastest, increasing as a share of all Ph.D.s to U.S. citizens from 15 percent in 1966 to 20 percent in 2000. In contrast, the share of all Ph.D.s in engineering and physical sciences fell over the period. As a whole, the share of Ph.D.s awarded in the science fields fell from 47 percent in 1966 to 39 percent in 2000. Among the

TABLE 9.1 Ph.D.s conferred by field.

Field	1966		2000	
	Number	Share of all Ph.D.s	Number	Share of all Ph.D.s
Sciences				
Engineering	1,690	11.3	2,206	7.9
Physical sciences	3,138	21.0	3,260	11.7
Life sciences	2,229	14.9	5,454	19.6
Total of sciences	**7,057**	**47.1**	**10,290**	**39.2**
Nonsciences				
Social sciences	2,268	15.1	5,848	21.0
Education	2,875	19.2	5,532	19.8
Humanities	1,987	13.3	3,096	11.1
Other	787	5.3	2,492	8.9
Total of nonsciences	**7,917**	**52.9**	**16,968**	**60.8**

Source: NSF, Survey of Earned Doctorates.

Note: For field classifications and examples of "sciences" see Appendix.

nonscience fields, social sciences increased the fastest, from 15 percent in 1966 to 20 percent in 2000, with education and humanities each having the same share of Ph.D.s in 1966 and 2000.

TYPE OF UNDERGRADUATE INSTITUTION

Next, we explore the composition of doctorate recipients by the type of institution they attended as undergraduates. The undergraduate origins of Ph.D. recipients are relevant for understanding the role different types of institutions play in preparing and motivating undergraduate students to pursue graduate study. The number of Ph.D. recipients who attended a certain type of college or university as undergraduates reflects a variety of factors. First, institutions that train large numbers of undergraduates (e.g., public universities) will produce a lot of Ph.D. recipients merely because of their scale, even if a relatively small share of their undergraduates pursue a Ph.D. Second, institutions may differ in the share of their undergraduates who pursue Ph.D. study merely because of the sorting (by academic ability and interests) created by the college admissions process. Third, institutions and their faculty may directly encourage their undergraduate students to pursue Ph.D. study. Of course, this could happen through strong preparation in a particular discipline. In addition, it might also be important to expose undergraduates to research, such as writing a senior thesis or assisting a professor with a research project.

Looking first at the control of the institution, the number of Ph.D. recipients who attended public institutions as undergraduates far exceeds the number who attended private institutions. For instance, 59 percent of Ph.D. recipients in 2000 attended public institutions. However, Ph.D. propensity is greater among students who attended private institutions: 2.8 percent in 2000 compared to 2.1 percent among students who attend public institutions. The share of Ph.D. recipients who attended public institutions as undergraduates has grown over time from 51 percent in 1963 to 59 percent in 2000. This growth reflects the differential growth in the size of public and private undergraduate sectors. In particular, the public bachelor's degree sector grew much more rapidly from 1963 to 1974 (see Groen and Rizzo 2003). The trends in Ph.D. propensity, however, are remarkably similar for students from both sectors and follow the overall pattern in Figure 9.4.

The majority of Ph.D. recipients attended research universities as undergraduates. In 1963 and 2000, 61 percent of Ph.D. recipients attended research universities. In addition, 21 percent of Ph.D. recipients in 2000 attended master's-level institutions, and 18 percent attended liberal arts colleges. Liberal arts colleges have played a slightly increasing role in preparing students for Ph.D. study, increasing their share by 5 percentage points over the 1963–2000 period. While liberal arts colleges produce the fewest Ph.D. recipients of the three groups, the propensity of their students to earn Ph.D.s is comparable to that of research universities. In 2000 Ph.D. propensity was about 3 percent for students from both types of institutions, compared to 1.5 percent among students from master's institutions. The time pattern of Ph.D. propensity for each group is similar to the overall pattern among U.S. citizens (Figure 9.5).

The relatively high Ph.D. propensity for students who attended private institutions and liberal arts colleges presumably reflects the tendency for these institutions to attract students who are academically talented. To determine the "quality" of an undergraduate institution, we use *America's Best Colleges* (2003) rankings to identify the top-ten liberal arts colleges and the top-ten research universities. Not surprisingly, Ph.D. propensity is much higher among students at top-ten institutions. Ph.D. propensity in 2000 was 13 percent among students at top-ten liberal arts colleges and 10.6 percent at top-ten research universities, compared to about 3 percent among all institutions in each category.

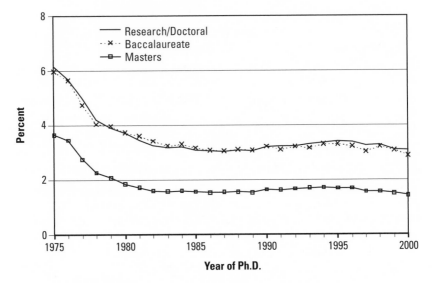

FIGURE 9.5 Ph.D. propensity by type of baccalaureate institution, 1975–2000.

Source: Ph.D. data from NSF, Survey of Earned Doctorates. Baccalaureate data from NCES, *Earned Degrees Conferred.*

Note: Institution categories defined using the 1994 Carnegie Classifications (Carnegie Foundation for the Advancement of Teaching 1994).

From 1975 to the mid-1980s, Ph.D. propensity at these institutions dropped dramatically, following the overall trend. Since the mid-1980s, Ph.D. propensity overall and within the broad categories defining institution types has been relatively stable. At top-ten institutions, by contrast, Ph.D. propensity rose over the period: from 1986 to 2000, it rose from 8.5 percent to 13 percent among liberal arts colleges and from 9.1 percent to 10.6 percent among research universities (Figure 9.6). The differential trend at top-ten institutions might reflect the increasing concentration of top students at elite institutions over this period (Cook and Frank 1993; Hoxby 1997).[8]

DEMOGRAPHIC CHARACTERISTICS

Gender

The presence of women among doctorate recipients has increased markedly since the 1960s. In 1966 men earned 88 percent of all Ph.D.s awarded to

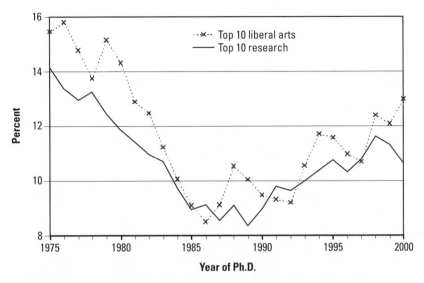

FIGURE 9.6 Ph.D. propensity for top-ten baccalaureate institutions, 1975–2000.

Source: Ph.D. data from NSF, Survey of Earned Doctorates. Baccalaureate data from NCES, *Earned Degrees Conferred.*

Note: Institution categories based on *America's Best Colleges* (2003) rankings.

U.S. citizens. By 2000 the gender gap in overall Ph.D.s awarded had nearly vanished, with men and women each earning roughly half of all Ph.D.s. The share of Ph.D.s awarded to women has increased continuously over the period in each of five broad fields (Table 9.2). In science fields, the share of Ph.D.s awarded to women increased from 6 percent in 1966 to 37 percent in 2000. While the gender gap has disappeared in the aggregate, women are still relatively more represented in social sciences and education and less represented in the sciences.

While the share of Ph.D.s awarded to women has increased continuously since the 1960s, the gender gap as measured by the raw difference in Ph.D.s awarded has not. It widened substantially in the late 1960s and early 1970s, reaching a peak in 1971 (Figure 9.7). This was due to differences between men and women in the time pattern of Ph.D.s awarded. Since 1963 the number of Ph.D.s awarded to women has increased steadily. However, the number of Ph.D.s awarded to men increased rapidly in the late 1960s, decreased from the early 1970s to the mid-1980s, and has been relatively stable since then. The time pattern for science fields is similar to that for all fields.

TABLE 9.2 Ph.D.s conferred to women as share of total.

Field	1966	1980	2000
Sciences	5.7	18.9	36.8
Social sciences	16.0	36.4	57.9
Education	18.7	46.1	66.1
Humanities	19.4	41.9	49.7
Other	13.7	31.2	47.1
Total	**12.0**	**33.1**	**49.4**

Source: NSF, Survey of Earned Doctorates.

The rise and fall in Ph.D.s awarded to men was driven in large part by variation in Ph.D. propensity across cohorts. In particular, in the late 1960s Ph.D. propensity for men increased from 6 percent in 1963 to 10 percent in 1971. However, it reversed course at this point and fell dramatically over the 1970s. Between the 1971 and 1981 Ph.D. cohorts, Ph.D. propensity for men fell from 10 percent to 3.2 percent (Figure 9.8). The abrupt reversal in Ph.D. propensity for men in 1971 appears to have been the result of the end of Vietnam War draft deferments for graduate students in 1967–68 (Bowen, Turner, and Witte 1992), just as the increased propensity during the 1960s reflected the availability of draft deferments for graduate study. Women were not subject to the draft, and the pattern of Ph.D. propensity over the early 1970s is different for them. From 1971 to 1975, Ph.D. propensity for women rose, continuing the trend from the early 1960s.

Over the late 1970s, however, Ph.D. propensity fell for women as well, albeit less rapidly than it was falling for men. This suggests a general weakening in Ph.D. prospects. Since 1980, Ph.D. propensity has been relatively stable for both groups, at about 3 percent for men and 2.5 percent for women. Despite the lower Ph.D. propensity for women, the tremendous growth in the number of bachelor's degrees awarded to women allowed the number of Ph.D.s awarded to women to continue to rise. For men, however, the number of Ph.D.s has been relatively stable since the early 1980s because both Ph.D. propensity and the number of bachelor's degrees awarded have been stable. Within science fields, Ph.D. propensity is higher for both men and women, but the time trends are similar to those over all fields.

Examining these trends in Ph.D. propensity raises questions about the likelihood of men and women in different cohorts to pursue post-bachelor's training generally, not just Ph.D.s. In particular, to what extent do changes in Ph.D. propensity over time reflect shifts between Ph.D.s and other

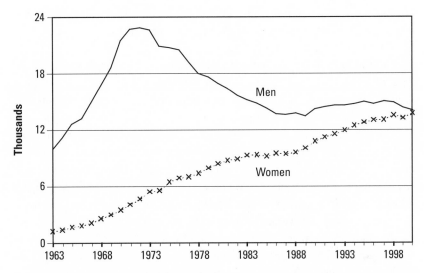

FIGURE 9.7 Ph.D.s conferred by gender, 1963–2000.

Source: NSF, Survey of Earned Doctorates.

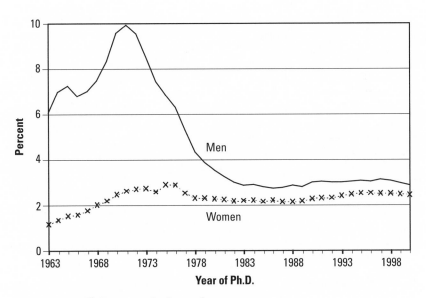

FIGURE 9.8 Ph.D. propensity by gender, 1963–2000.

Source: Ph.D. data from NSF, Survey of Earned Doctorates. Baccalaureate data from NCES, *Earned Degrees Conferred.*

graduate degrees, versus attainment of graduate degrees altogether? We address this question with data on degrees awarded to men and women since 1971 in three professional fields: medicine, law, and business (M.D., J.D., and M.B.A. degrees). Professional degrees in these fields are the principal alternatives to the Ph.D. for most students. Compared to Ph.D. programs, professional degree programs typically offer higher completion rates and shorter times to degree.

For each of the three fields, we examine trends in both the number of degrees conferred and the propensity of men and women with bachelor's degrees to earn the professional degree (Figure 9.9). We measure propensity following the method for Ph.D.s, except that we use a five-year lag to reflect the shorter length of these programs relative to Ph.D. programs. The trends reveal several interesting facts. First, the falling Ph.D. propensity for men in the 1970s is not explained by shifts from Ph.D.s to professional degrees. The propensity of men to earn professional degrees fell in medicine and law (after 1973) and was relatively stable in business. The declines in M.D. and J.D. propensities is consistent with the explanation that Ph.D. propensity fell in the 1970s because of the end of the Vietnam

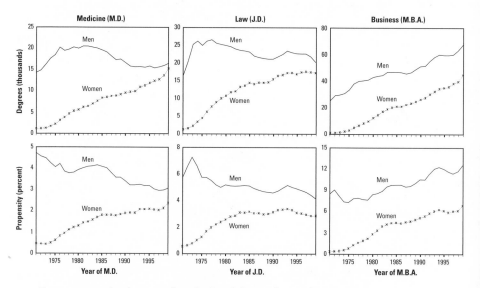

FIGURE 9.9 Attainment of professional degrees by gender, 1971–1999.

Source: NCES, Earned Degrees Conferred.

Note: "Propensity" is the number of professional degrees conferred in a year divided by the number of bachelor's degrees conferred five years earlier.

War. Second, while Ph.D. propensity has been stable for men and women since the early 1980s, M.B.A. propensity has increased for both men and women, J.D. propensity has been stable for both men and women, and M.D. propensity has decreased for men and increased for women. Third, the share of men with bachelor's degrees going on to earn either a Ph.D. or a professional degree decreased in the 1970s and increased in the 1980s and 1990s. For women, however, the corresponding share increased in all three decades.

Race and Ethnicity

It is well known that certain racial and ethnic minorities are underrepresented among doctorate recipients relative to their presence among bachelor's degree recipients or among the general population. Indeed, this is the basis for special assistance at the undergraduate level to encourage minorities to pursue Ph.D. programs, especially in the sciences. Among Ph.D.s awarded to U.S. citizens in 2000, 5.8 percent were awarded to blacks, 4.9 percent to Asians, and 4.2 percent to Hispanics. Since 1975 the share of Ph.D.s awarded to each minority group has increased (Figure 9.10). In particular, the share of Ph.D.s awarded to Asians and Hispanics increased steadily from around 1 percent in 1975 to 4–5 percent in 2000. The share

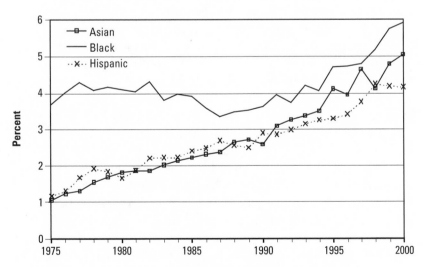

FIGURE 9.10 Ph.D.s conferred to racial and ethnic minorities as share of total, 1975–2000.

Source: NSF, Survey of Earned Doctorates.

TABLE 9.3 Ph.D.s conferred to racial and ethnic minorities as share of total.

Field	Black			Hispanic			Asian		
	1990	2000	Diff.	1990	2000	Diff.	1990	2000	Diff.
Sciences	1.3	3.1	1.8	2.3	3.5	1.2	4.2	7.5	3.3
Social sciences	3.8	6.3	2.5	3.6	5.0	1.4	1.9	3.8	1.9
Education	8.1	12.2	4.1	3.2	4.9	1.7	1.2	2.2	1.0
Humanities	2.6	3.6	1.0	4.3	4.8	0.5	1.0	2.9	1.9
Other	3.2	5.1	1.9	2.0	3.0	1.0	1.7	4.1	2.4
Total	3.6	5.8	2.2	2.9	4.2	1.3	2.6	4.9	2.3

Source: NSF, Survey of Earned Doctorates.

awarded to blacks, however, fluctuated between 3 and 4 percent from 1974 to the late 1980s before increasing in the 1990s.

Given the increasing presence of minority groups among Ph.D.s awarded in the 1990s, it is relevant to ask whether the trends are different by field of study. To be sure, the presence of minority groups varies greatly across fields at any particular point in time. In 2000, for instance, blacks represented 12 percent of Ph.D.s awarded in the field of education but only 3 percent of Ph.D.s awarded in the sciences (Table 9.3), while Asians represented 2 percent of Ph.D.s in education and 8 percent of Ph.D.s in the sciences. The presence of Hispanics is the least variable of the three groups across fields.

Within science fields, the time pattern for the share of Ph.D.s awarded to minority groups has followed the trend over all fields. Notably, in the 1990s each group experienced an increase in its share of science Ph.D.s. In fact, this happened not only in science but also in education, humanities, and social sciences (Table 9.3). Blacks and Asians experienced the largest gains in their share of Ph.D.s in the fields in which they had the strongest presence in 1990: blacks in education and Asians in the sciences. Hispanics made the largest gains in education.

Increasing shares of Ph.D.s awarded to minority groups could reflect demographic trends and/or an increasing propensity of bachelor's degree recipients to earn Ph.D.s. For doctorate recipients in 2000, Ph.D. propensity was 3.4 percent for Asians, 2.5 percent for blacks, 2.3 percent for Hispanics, and 2.5 percent for whites. (In the sciences, Ph.D. propensity was 4.8 percent for Asians, 2.3 percent for blacks, 3.2 percent for Hispanics, and 5.2 percent for whites.) Trends in Ph.D. propensity by race from 1986 to 2000 reveal some interesting facts (Figure 9.11). Over this period, Ph.D. propensity increased modestly for blacks, increased slightly for Hispanics, and

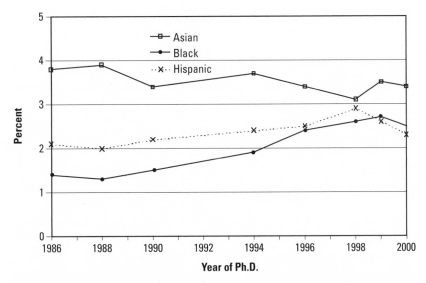

FIGURE 9.11   Ph.D. propensity by racial and ethnic group, 1986–2000 (selected years).

Source: Ph.D. data from NSF, Survey of Earned Doctorates. Baccalaureate data from NCES, *Earned Degrees Conferred.*

Notes:
Due to limitations in the baccalaureate data, measures of Ph.D. propensity can be constructed
   for selected years only (even-numbered years [except 1992] and 1999).
Ph.D. propensity for whites is similar to Ph.D. propensity over all races (see Figure 9.4).

decreased slightly for Asians. As a consequence of these patterns (and the relatively stable propensity for whites), racial differences in Ph.D. propensity narrowed from 1986 to 2000. (Trends within the sciences are similar.)

Since Ph.D. propensity decreased for Asians, their increasing share of Ph.D.s must be due to increases in the number of bachelor's degrees awarded. Yet for blacks and Hispanics Ph.D. propensity increased. Therefore, the increasing share of Ph.D.s awarded to blacks and Hispanics is due to increases in both Ph.D. propensity and in the number of bachelor's degrees awarded.

CONCLUSION

We have identified several key trends in the composition of doctorates awarded to U.S. citizens since the early 1960s. First, the propensity of bachelor's degree recipients to earn Ph.D.s varied widely during the 1960s and

1970s, especially for men. In particular, Ph.D. propensity for men declined rapidly in the early 1970s. This change appears to have been driven in large part by the end of Vietnam War draft deferments for graduate students. More generally, we would like to better understand the factors behind the trends in Ph.D. propensity over the entire period. For instance, how does Ph.D. propensity depend on factors reflecting the cost of graduate study, such as tuition, the availability of scholarships and research grants, and the foregone earnings while in graduate school? How does it depend on the structure of Ph.D. programs and the wages and employment opportunities in occupations requiring a Ph.D. and those in alternative careers? Evidence on these questions would appear to be important in evaluating the effectiveness of various governmental and institutional policies.

Regarding the undergraduate origins of doctorate recipients, the number of Ph.D. recipients who attended particular types of institutions depends on both the relative size of the sectors and the propensity of students from each sector to earn Ph.D.s. The time trends in Ph.D. propensity are remarkably similar across different sectors. What stands out, instead, are differences in Ph.D. propensity across sectors at a particular point in time. For instance, students who attended private institutions and research universities as undergraduates are more likely to earn Ph.D.s. An open question is the extent to which these differences reflect the academic qualifications of students on entering undergraduate training versus the ability of institutions to create interest and aptitude for graduate study.

In terms of demographics, the most significant change over the past forty years has been the dramatic increase in the presence of women among doctorate recipients. Overall, Ph.D.s awarded to women as a share of the total increased from 11 percent in 1963 to 49 percent in 2000. This increase was driven largely by increases in the number of women earning bachelor's degrees and by decreases in Ph.D. propensity for men. Over the 1980s and 1990s, professional degrees (especially in business) have become increasingly common for both men and women, while Ph.D. propensities have remained stable. Separately, racial and ethnic minorities have increased their presence among doctorate recipients within several fields, especially since 1990. Differences in Ph.D. propensity across racial and ethnic groups have narrowed since the mid-1980s.

APPENDIX: Field Classification

Sciences	Nonsciences

Physical Sciences
   Chemistry
   Physics and Astronomy
   Earth, Atmospheric, and Marine
     Sciences
   Mathematics and Statistics
   Computer Sciences

Life Sciences
   Biological Sciences
   Agricultural Sciences
   Health and Medical Sciences
   Other Life Sciences

Engineering
   Aerospace Engineering
   Chemical Engineering
   Civil Engineering
   Electrical Engineering
   Mechanical Engineering
   Materials Engineering
   Industrial Engineering
   Other Engineering

Humanities
   History
   English and Literature
   Foreign Languages
   Other Humanities

Social Sciences
   Psychology
   Economics
   Political Science and Public
     Administration
   Sociology
   Anthropology
   Linguistics
   Other Social Sciences

Other
   Business and Management
   Arts and Music
   Religion and Theology
   Communication and Librarianship
   Social Service Professions
   Other Professional Fields
   Other Fields

Notes

This article is a condensed version of the paper we presented at the "Science and the University" conference in May 2003 at Cornell University. We are grateful to the Andrew W. Mellon Foundation and the Atlantic Philanthropies (Inc.) USA for funding through the Cornell Higher Education Research Institute. We thank Ronald Ehrenberg, Walter Cohen, and conference participants for their helpful comments and Andrew Nutting for his research assistance.

1. Recent studies using similar data sources include Bowen and Rudenstine (1992), Ehrenberg (1992), and Lomperis (1992).

2. For convenience, we use the terms "Ph.D." and "doctorate" interchangeably, since most research doctorates are Ph.D.s. Among doctorates that are not Ph.D.s, the large majority are in the field of education (Ed.D.).

3. The overall response rate for the 2000 survey was 92 percent. Annual response rates have varied over the 1967–2000 period from 91 percent to 98 percent (Hoffer et al. 2001).

4. Our data from the Survey of Earned Doctorates for 1966–2000 are taken

from WebCASPAR, a Web-based database system maintained by the National Science Foundation containing information from a variety of surveys. Data for 1963–65 are taken from a published report (National Academy of Sciences 1967).

5. Throughout the chapter, year t refers to the academic year ending on June 30 of calendar year t. For example, "year 2000" refers to the period from July 1, 1999, to June 30, 2000.

6. Data for bachelor's degrees awarded 1966–91 are taken from WebCASPAR (see note 4). For 1954–65, our bachelor's degree data are based on published reports (National Center for Education Statistics [various years]).

7. An alternative would be to organize the Survey of Earned Doctorates data by bachelor's degree cohort rather than Ph.D. cohort, as in Bowen, Turner, and Witte (1992). However, the underlying micro data are not publicly available and the public-use tabulations are organized by Ph.D. cohort. As a consequence, researchers typically organize the data by Ph.D. cohort.

8. Cook and Frank (1993) present evidence that the concentration of top students increased slightly between the 1960s and 1970s but increased substantially between the 1970s and 1980s. The increase in concentration in the 1980s roughly corresponds to the timing of the increase in Ph.D. propensity among students at top-ten baccalaureate institutions.

10

Where Do New U.S.-Trained Science and Engineering Ph.D.s Come From?

RICHARD B. FREEMAN, EMILY JIN, AND CHIA-YU SHEN

Have the demographics of Ph.D. graduates changed from those in earlier decades? Have highly selective Ph.D.-producing universities maintained their share of science and engineering (S&E) Ph.D.s or have newer institutions and programs increased their proportion of doctorates? Do the majority of S&E Ph.D.s matriculate at the most prestigious undergraduate colleges and universities or do Ph.D. graduates come from a wider set of baccalaureate programs? Has there been a shift in these patterns over time?

This paper uses data from the Survey of Earned Doctorates (SED) to examine questions such as these concerning the origins of newly U.S.-trained S&E Ph.D.s in 2000 compared to earlier decades. The data document a significant shift in the demographic origins of U.S.-trained S&E Ph.D.s from U.S.-born white males to women, minorities, and the foreign born. They also show that the proportion of Ph.D.s from less selective doctoral programs and universities has increased noticeably. In addition, the number of bachelor's graduates earning science and engineering Ph.D.s from the most selective liberal arts colleges increased modestly relative to the number earning Ph.D.s from the selective research universities. Still, both the number and proportion of bachelor's graduates obtaining S&E Ph.D.s are bigger at the research universities.

The New Demography of U.S.-Trained Science and Engineering Ph.D.s

Ph.D. graduates from U.S. universities in the early 2000s differ markedly from graduates three or four decades earlier. In the 1960s and early 1970s, the vast majority of Ph.D.s graduating in science and engineering from U.S. universities were native-born white males. In 2000, native-born white males were a distinct minority. Figure 10.1 shows the decline in the native-born male share of Ph.D.s. In 1966, 71 percent of Ph.D. graduates were U.S.-born males;[1] 6 percent were U.S.-born females; and 22.5 percent were foreign-born students. In 2000, 36 percent of Ph.D. graduates were U.S.-born males; 25.4 percent were U.S.-born females; and 38.9 percent were foreign born. In addition, the graduates obtaining S&E Ph.D.s in 2000 were older than earlier cohorts of Ph.D.s. In 1966, the median age of an S&E Ph.D. was thirty. In 2000 the median age of an S&E Ph.D. was thirty-two. Measured by mean ages, the aging of new graduates was greater: in 1966 the mean age of a Ph.D. graduate was 31.5 years; in 2000 the mean age was 33.9.

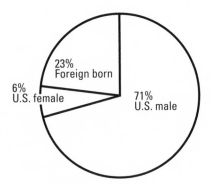

FIGURE 10.1a Changing demographics of U.S.-trained Ph.D.s, 1966.

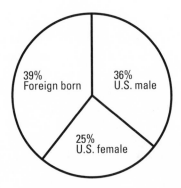

FIGURE 10.1b Percentage distribution of Ph.D.s by demographic characteristics, 2000.

Figure 10.2 shows that the foreign-born proportion of U.S.-trained Ph.D.s was about 21 percent in the late 1960s; it then rose gradually through 1980 and then increased sharply through the mid-1990s, leveling off at about 40 percent. This is consistent with the evidence by Black and Stephan in chapter 6 that the increase in the share of U.S. Ph.D.s awarded to temporary residents occurred discontinuously. The increase in the number of degrees awarded to foreign-born students, regardless of their citizenship status at the time the degree was awarded, tracks well the increase in the number of degrees by citizenship status. Disaggregated by field, the data show that the foreign born have become a majority of graduates in some disciplines, including branches of engineering, while increasing in all fields.

To see the degree to which foreign-born Ph.D.s come from U.S. baccalaureate backgrounds rather than overseas baccalaureate programs, we tabulated the location of undergraduate degrees of foreign-born Ph.D.s. The statistics show that the vast majority of foreign-born Ph.D.s obtained foreign bachelor's degrees. The late 1980s–early 1990s jump in the foreign-born share of Ph.D.s, in particular, came primarily from students educated overseas. Still, there was a substantial increase in the proportion of U.S. science and engineering Ph.D.s granted to foreign-born students with a U.S. bachelor's education. This reflects in part the large inflow of immigrants with children in the late 1970s and 1980s. In 1958–68, 6.3 percent of

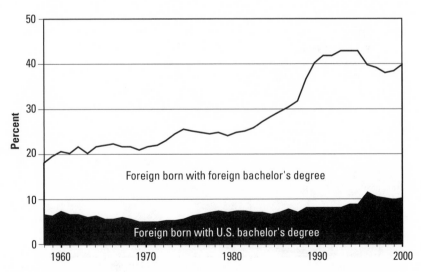

FIGURE 10.2 U.S. bachelor's degrees and foreign bachelor's degrees of foreign-born Ph.D.s, 1958–2000.

all U.S. Ph.D.s were foreign born with an undergraduate degree from a U.S. institution. During the period 1991–2001, the percentage of U.S. Ph.D.s who were foreign born with a U.S. undergraduate degree had increased to 9.4 percent. The increased share of foreign-born Ph.D.s with foreign bacca-laureates was, however, bigger: from 14.8 percent of all Ph.D.s in 1958–68 to 31.7 percent of all Ph.D.s in 1991–2001.

Among U.S. citizens, the proportion of Ph.D.s granted to minorities—Asian and Pacific Islanders, blacks, and Hispanics—has trended upward. Figure 10.3 displays the increased proportion of science and engineering Ph.D.s awarded U.S. women and minorities over time. It records the share of U.S.-citizen Ph.D.s that went to women from 1966 to 2000 and the share of U.S.-citizen Ph.D.s that went to minorities from 1973 (when the surveys began recording detailed minority representation) to 2000. The figure shows steady increases in the female and minority shares of U.S.-citizen S&E Ph.D.s beginning in the 1970s proceeding through 2000, with only the barest sign of leveling-off by the end of the period. The female share of U.S.-citizen Ph.D.s rose from below 10 percent in the late 1960s to over 40 percent by 2000. The share of U.S.-citizen S&E Ph.D.s going to Asian and Pacific Islanders was negligible in 1973 but increased to 6.3 percent in 2000—above their proportion of the U.S. population. Black and other minorities (American Indian and Hispanics) increased their share of U.S.-citizen Ph.D.s

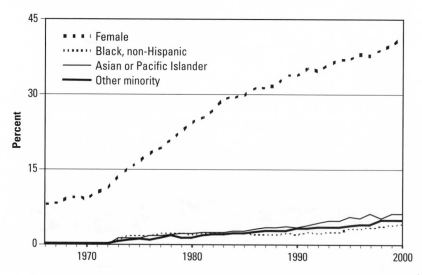

FIGURE 10.3 Women or minorities' share of U.S. citizen Ph.D.s, 1966–2000.

to approximately 2.7 percent in 2000, compared to negligible numbers in 1973, though these proportions are below their shares of the population.

The increased female and minority shares of U.S.-citizen S&E Ph.D.s occurred in part because of increases in the number of Ph.D.s granted these traditionally underrepresented groups. The number of U.S. females gaining Ph.D.s increased from 748 in 1966 to 2,110 in 1973 to 6,614 in 2000. The number of U.S.-citizen minority Ph.D.s increased from 450 in 1973 to 2,374 in 2000. But the increased female and minority shares of U.S.-citizen S&E Ph.D.s also reflects a 37 percent drop in the number of U.S.-citizen white males gaining Ph.D.s from 1973, when 12,518 white men earned S&E Ph.D.s, through 2000, when 7,829 white men earned S&E Ph.D.s.[2]

Interpretation

The demography of Ph.D. production reflects the decisions of different groups: students, Ph.D.-granting institutions, and undergraduate institutions, all influenced by government stipend and research and development (R&D) policies and by the labor market for scientists and engineers and other high-level occupations.

On the student side, the changed demographic composition of Ph.D. graduates, especially the increased ratio of the foreign-born to domestic students, reflects the different economic incentives that face different groups of students in earning Ph.D.s compared to other career alternatives. Foreign-born students, particularly those with foreign bachelor's degrees, can often earn much more with a U.S. doctorate than they could earn from working in alternative careers in their native country, in part because the S&E doctorate opens the door to working in the United States or working for U.S. and other multinational firms in some other locale. By way of contrast, U.S.-citizen students have more alternatives, such as attending medical school, law school, or business school, which gives them access to the high earnings available to educated people in the United States. They can also work as a scientist or engineer without obtaining a Ph.D. As a result, they have less incentive to invest in a science and engineering Ph.D. than do otherwise comparable foreign-born bachelor's graduates. In addition, the foreign born often receive financial support for attending graduate school in the form of graduate research and teaching assistantships. These stipends can seem quite large when compared with what they could receive in their native country. Most important, the large increase that has occurred in the number of young persons earning bachelor's degree overseas has increased the potential supply of bright students to science and engineering in the United States.

Why have women and minorities increased their share of native-born Ph.D. graduates? One potential factor, stressed by Groen and Rizzo in chapter 9, is that women and minorities have increased their share of bachelor's science and engineering degrees. This is a correct observation, and it identifies the increased number of women and minorities obtaining S&E degrees as the critical factor. But it does not explain why women and minorities have increasingly chosen to major in science and engineering at the bachelor's level or why this increased number has barely affected the proportion of female and minority bachelor's graduates choosing to continue on toward the Ph.D. Why should these groups go into S&E when the job market discouraged white males from doing so? I interpret the influx of women and minorities as a "correction" to their previously exceptionally low representation in science and engineering. Groups that had been largely excluded from S&E work would not need improved labor market conditions to induce them to enter the fields. If this interpretation is correct, as the female and minority proportions in science and engineering approach their proportions in the population, their supply behavior should resemble more closely that of their male counterparts.

Who Produces Ph.D.s?

In 1970, 214 universities/campuses granted Ph.D.s in science and engineering. In 2000, the number had grown to 339. Between 1960 and 2000, the U.S. university sector "produced" an increasing number of Ph.D.s in science and engineering. There were 6,520 S&E Ph.D.s in 1960; 18,052 in 1970; 17,775 in 1980; 22,868 in 1990; and 25,951 in 2000. A priori, the increased number of Ph.D.s could have come from increased Ph.D. production at the larger leading universities that traditionally have produced 400–500 S&E Ph.D.s a year, such as University of California, Berkeley, University of Illinois at Urbana-Champaign, University of Wisconsin–Madison, University of Michigan, University of Minnesota, Massachusetts Institute of Technology (MIT), and Stanford University. It could also have come from proportional increases in degrees among all Ph.D.-granting universities regardless of size, quality, or vintage. Or it could have come from increases in the number of degrees from universities with smaller programs (such as Providence College, University of the Pacific, University of California, Santa Cruz, or Ball State University) or from less prestigious programs or from newly formed programs. Our analysis shows that the growth of S&E Ph.D.s in the United States was fueled by an expansion of degrees from smaller, lower quality institutions rather than from increased numbers of graduates from elite universities.

Table 10.1 presents the data for this conclusion. It shows how the distribution of science and engineering Ph.D.s changed over time along the specified dimensions. The columns "by number of Ph.D.s" give the distribution of degrees by the number of Ph.D.s granted. They show that the proportion of Ph.D.s granted by the ten largest Ph.D.-granting institutions fell from 24.11 percent in 1970 to 16.68 percent in 2000, the proportion of

TABLE 10.1 Concentration of Ph.D.s by doctoral institution, 1970–2000.

Year	Total Ph.D.s	By number of Ph.D.s Percent top			HHI*	By Carnegie Classification		
		10	11–25	26–40		Research	Doctoral	Other
1970	18052	24.11	21.64	13.38	122.98	91.6	7.78	0.62
1975	18799	21.30	19.87	13.67	105.51	88.82	9.35	1.83
1980	17775	20.19	18.82	13.47	98.27	86.17	10.58	3.25
1985	18888	20.64	17.95	13.41	96.99	86.68	10.28	3.03
1990	22868	19.62	18.29	12.70	91.32	84.22	11.45	4.32
1995	26536	18.26	18.57	12.63	86.54	83.31	11.66	5.03
2000	25951	16.68	17.51	12.50	78.69	81.39	12.50	6.11

Year	By number of top Ph.D. programs (quality)				By federal R&D spending (percent of Ph.D.s granted by institution over total Ph.D.s)		
	10+	3–10	1–2	0	% top 10	% top 20	% top 40
1970	26.11	15.36	20.98	37.55	20.17	37.52	51.25
1975	23.66	14.83	20.90	40.61	19.01	37.99	49.50
1980	22.37	14.17	21.19	42.27	16.88	35.06	45.88
1985	22.04	14.43	20.44	43.10	18.60	34.52	47.59
1990	21.03	14.22	19.76	44.99	17.66	33.83	45.26
1995	19.63	14.48	20.33	45.56	15.60	32.99	45.42
2000	18.25	13.69	19.47	48.59	13.09	30.92	42.35

Source: NSF, Survey of Earned Doctorates, various years. Data available on WebCASPAR.

Notes:

In classifying the institutions by federal financed R&D expenditure, I used the 1972 ranking for the year 1970 because 1970 has no data on federal R&D expenditure. The percentages are obtained, however, using 1970's number of Ph.D.s.

We eliminated Woods Hole Oceanographic Institute from the calculation of the percentage of top 26–40 institutions according to federal financed R&D expenditure since from 1970 to 1995 Woods Hole had no Ph.D. recipients; we included the 41st institution in terms of federally funded R&D instead.

* HHI = Herfindahl-Hirschman Index.

the next largest fifteen Ph.D. producers fell from 21.64 percent to 17.51 percent, and the proportion of the next 15 largest producers fell by 0.88 percentage points. As a summary of the overall concentration of S&E Ph.D.s among large Ph.D.-producing universities, we calculated the Herfindahl-Hirschman index of concentration (HHI in Table 10.1). This index is the sum of squares of the proportions of degrees given by each degree-granting institution. It is a standard measure of concentration in industrial organization, used, for instance, by the U.S. Department of Justice.[3] Analysts of industrial organization consider markets in which the index is between 100 and 180 to be moderately concentrated and those in which the index exceeds 180 to be concentrated. From this perspective, the drop in the index for Ph.D.-granting universities from 122.98 to 78.69 implies that Ph.D. production went from moderately concentrated to highly competitive.

The columns labeled "By Carnegie classification" in Table 10.1 use the Carnegie Foundation's well-established taxonomy of higher education institutions[4] to divide the institutions into three groups: leading research universities; doctoral universities that grant at least twenty degrees in any field or ten or more in three fields but that did not fit the criterion for being a research university; and all other institutions. The data shows a drop in the proportion of degrees from the research universities and corresponding increases in the proportion of degrees from doctoral universities and other institutions.

The next set of columns in Table 10.1, labeled "By number of top Ph.D. programs (quality)," classifies universities by the number of Ph.D. programs that attained high rankings in the 1982 National Research Council (NRC) rankings. These columns show a decrease in the share of Ph.D. production in universities that the NRC rated as having ten or more high-quality programs, a decrease in the share that the NRC rated as having three to ten high-quality programs, and a commensurate increase in the proportion of S&E Ph.D.s from universities without NRC-rated high-quality programs.

Finally, the last set of columns in Table 10.1 record the proportion of Ph.D.s coming from universities in the top ten receivers of federal R&D money. The share of Ph.D.s from these institutions fell sharply from 1985 to 2000 as Ph.D. production expanded. The number of S&E Ph.D.s from the top federal R&D recipients in 2000 was lower than the number from the top federal R&D recipients in 1985.

In sum, the increase in science and engineering Ph.D.s in the United States largely took the form of expansion of smaller and less prestigious programs. The absolute number of Ph.D.s awarded by the very top programs remained almost constant during the period. More than two-thirds of the 14,381 increase in the number of Ph.D.s granted in science and engineering between 1966 and 2000 occurred among institutions ranked below the top forty Ph.D. producers. Eighty-one percent of the growth occurred among universities with few high-quality programs.[5] Expansion of production in smaller institutions with few high-quality programs contrasts with the pattern of growth in many other sectors of the economy, where increases in production typically come from large firms (automobiles, steel, software). What might explain the concentration of growth in smaller Ph.D.-granting institutions with few high-quality programs?

The simplest economic story that explains why the higher education sector expanded Ph.D. production by developing new or small Ph.D. programs rather than by increasing the numbers of Ph.D.s from major research universities is that the cost schedule for producing Ph.D.s is very rigid due to capacity constraints set by faculty, plant, or other characteristics of existing programs. This contrasts with potential increasing returns to scale (at least up to much larger production) in many other sectors. Expansion of Ph.D. production through new or smaller doctorate programs may also reflect the U.S. system of financing higher education. In the public sector, legislatures have been willing to fund new Ph.D. programs in their state universities or bestow university status on institutions that were formally called colleges, enhancing the ability of the institution to offer a Ph.D. Sometimes this has been facilitated by the availability of "earmarked money" (see de Figueiredo and Silverman herein, chapter 2). Local citizens are also likely to lobby the legislature for such change, seeing new Ph.D. programs as a means of bestowing status on their communities. Growth in the number of Ph.D. programs also was facilitated by aggressive hiring in the late 1960s and early 1970s by new universities. Many of these universities were able to hire faculty who in earlier years would have gone to more highly ranked institutions. Once hired, the new faculty worked to create Ph.D. programs at their institutions (Stephan and Levin 1992).

WHO GETS PH.D.s WHERE?

Have the growing number of foreign-born and female Ph.D.s obtained their degrees in the same doctoral institutions as U.S.-born men, or are they disproportionately represented in the smaller or less prestigious universities? To answer this question, we calculated the distribution of Ph.D.s among U.S.-born men, foreign-born persons, and U.S.-born women at universities with three or more top-rated Ph.D. programs and at universities included in the top forty Ph.D.-granting programs in term of degrees granted.[6] We then compared the representation of these groups at these universities to their representation among all Ph.D.s in 1973 and 2000.

Table 10.2 summarizes the results of this analysis. The rows labeled "Three or more top programs" and "Top 40 Ph.D. producers" give the number of Ph.D.s in the specified university categories. The percentage figures record the percentages earned by the groups in the relevant university category. Thus, the 62.35 percent for U.S. men in the universities with three or more top programs in 1973 shows that U.S. men made up 62.35 percent of all Ph.D.s in universities with three or more top programs; while the 36.93 percent for U.S. men in 2000 shows that U.S. men made up 36.93 percent of all Ph.D.s in that category in 2000. The column labeled "Ratio" gives the ratio of a group's share of degrees in the relevant category relative to its share of all Ph.D.s in science and engineering. When these ratios are greater than one, the group is more represented in that university category than it is among all Ph.D.s. When a group's relative representation ratio increases, the group has obtained proportionately more Ph.D.s in

TABLE 10.2 Doctorate recipients in specified Ph.D.-granting institutions by demographic characteristic of Ph.D. recipient, 1973 and 2000.

Institutional characteristics	Demographic group	1973			2000		
		Number	Percent	Ratio*	Number	Percent	Ratio*
Three or more top programs (quality)	U.S.-born men	4694	62.35	0.96	3061	36.93	1.09
	Foreign-born	2049	27.22	1.10	3423	41.30	1.06
	U.S.-born women	785	10.43	0.99	1790	21.59	0.85
	All	7528	100.00		8289	100.00	
Top 40 Ph.D. Producers	U.S.-born men	6817	63.11	0.97	4474	36.92	1.04
	Foreign-born	2881	26.67	1.08	4972	41.03	1.05
	U.S.-born women	1104	10.22	0.97	2649	21.86	0.86
	All	10802	100.00		12117	99.99	
Total Ph.D.s in U.S.	U.S.-born men	12561	64.84	1.00	9225	35.55	1.00
	Foreign-born	4786	24.70	1.00	10087	38.87	1.00
	U.S.-born women	2026	10.46	1.00	6584	25.37	1.00
	All	19373	100.00		25951	100.00	

Source: NSF, Survey of Earned Doctorates, various years. Data available on WebCASPAR.

Note: In 2000 55 persons were of unknown sex among the U.S. born. This was 0.21% of the total.

*Ratio of percentage in group to percentage in total.

that category than in the past. Conversely, decreases in the relative representation ratio over time show that the group lost representation in that category.

The table shows that the 1973–2000 decline in the number of U.S. men obtaining S&E Ph.D.s was associated with an increase in their proportion in universities with three or more high-quality Ph.D. programs and an increase in their proportion in large Ph.D.-producing universities. This means that the drop in male S&E Ph.D.s occurred disproportionately at less prestigious, smaller Ph.D.-granting institutions. For foreign-born students the relative representation ratios in universities with three or more high-quality Ph.D. programs and in large Ph.D.-producing universities exceeded 1.00 in 1973 and remained above 1.00 in 2000, which implies that the huge increase in their share of Ph.D.s occurred with little fall in the quality of universities where they earned their Ph.D.s.[7]

The situation for women is quite different. As their share of S&E Ph.D.s increased, their relative representation in high-quality and large Ph.D.-granting universities fell. In 1973 women were slightly less likely to obtain a Ph.D. at these universities than other demographic groups. In 2000 women were much less likely to obtain a Ph.D. in the higher quality and larger universities than were other demographic groups. There are several possible explanations for why the increase in female S&E Ph.D.s occurred disproportionately at less prestigious, smaller Ph.D.-granting institutions. One possibility is that women enrolled in smaller, newer Ph.D. programs because those programs specialized in biological sciences, the social sciences, and psychology—fields that attract relatively many women. Another possibility is that women were more geographically limited in their choice of Ph.D. programs than men or the foreign born because of family issues. It is also possible, however, that women had difficulty gaining admission to the most prestigious and larger programs. We have not examined these or other possible causes for the disproportionate growth of female S&E Ph.D.s in the lower-quality and smaller universities.

Baccalaureate Origins of Ph.D.s

Some colleges and universities produce relatively large numbers of students who go on to get Ph.D.s. Among the S&E Ph.D.s who graduated in 2000, for example, 292 had an undergraduate degree from UC–Berkeley, 266 from Cornell University, 252 from the University of Michigan, 225 from the University of Illinois, 205 from MIT, and 198 from Pennsylvania

State University. These research universities combine large student bodies with high-quality undergraduate S&E programs. Relative to the number of students who matriculate, however, some small liberal arts colleges have produced more future doctorates than the major research universities. For example, upwards of 3–5 percent of graduates from highly selective liberal arts colleges such as Oberlin, Swarthmore, and Reed obtain Ph.D.s in science and engineering compared to 1–2 percent of bachelor's graduates from major research universities.

Have the baccalaureate origins of science and engineering Ph.D. recipients changed over time as the number of Ph.D.s, the demographic characteristics of Ph.D.s, and the composition of Ph.D.-granting institutions changed? To answer this question, we examined National Science Foundation (NSF) data on the U.S. baccalaureate origin of Ph.D.s along three dimensions.[8] First, we calculated the proportion of Ph.D.s earned by graduates from institutions that produce relatively many Ph.D.s. This statistic highlights the importance of the large research institutions in the baccalaureate origins of S&E doctorates noted earlier. Second, we calculated the proportion of Ph.D.s earned by graduates according to the Carnegie classification of colleges and universities, which distinguishes the major research universities from doctorate-granting institutions and primarily bachelor's-granting institutions.[9] This statistic provides insight into the extent to which institutions outside the research nexus send students to Ph.D. programs. Third, we calculated the percentage of Ph.D.s with origins in undergraduate programs by the selectivity of the bachelor's institution, as measured by Barron's guide to the selectivity of colleges. This categorization differentiates between the most competitive (or selective) undergraduate institutions (forty-five), highly competitive institutions (eighty-seven), very competitive institutions (forty), and all other institutions.[10]

Table 10.3a and Table 10.3b summarize the results of our analysis for the period 1970–2000. In Table 10.3a the columns under "Number of Ph.D.s by U.S.-baccalaureate and foreign-baccalaureate origin" show the number of Ph.D.s granted to U.S. bachelor's graduates as opposed to those with foreign bachelor's origins (and those for whom the bachelor's origin data is missing). These columns show the increase in the number of S&E Ph.D.s with foreign baccalaureates and an increase in the number with institutions unknown as well.

The columns under "Percentage of U.S. baccalaureate–origin Ph.D.s by number of Ph.D.s from baccalaureate institution" give the distribution of Ph.D.s by bachelor's institutions grouped by number of Ph.D.s with those

TABLE 10.3a Distribution of science and engineering Ph.D.s by U.S. baccalaureate–origin institution, 1970–2000.

Year	Number of Ph.D.s by U.S. Bacc.- and foreign Bacc.-origin				Percentage of U.S. Bacc.-origin Ph.D.s by number of Ph.D.s from Bacc. institution				
	Total from all Bacc. institutions	U.S. Bacc. origin	Foreign Bacc. origin	Blank Bacc.	Top 10	Top 25	Top 40	Top 175	Herfindahl-Hirschman Index
1970	18052	14898	3062	92	15.13	27.86	36.65	72.84	52.69
1975	18799	14845	3657	297	13.42	25.13	32.69	68.51	44.83
1980	17775	14057	3225	493	13.3	24.24	32.42	68.97	44.24
1985	18888	13699	4486	703	12.42	23.78	31.72	67.58	41.81
1990	22868	14739	7080	1049	12.95	24.15	32	67.9	42.84
1995	26536	15868	9478	1190	12.85	24.43	33.29	66.91	43.64
2000	25951	15677	7826	2448	13.14	25.36	34.43	67.9	45.13

Source: NSF, Survey of Earned Doctorates, various years. Data available on WebCASPAR.

baccalaureate origins. In these calculations we credit each bachelor's institution with the number of Ph.D. graduates from its school in the specified year. For instance, in 2000, 159 persons with a Harvard University bachelor's degree (from earlier years) gained a Ph.D., so we credit Harvard with 159 Ph.D.s. The data show a modest drop in the proportion of Ph.D.s granted to bachelor's from the largest undergraduate origin institutions between 1970 and 1975 but considerable stability thereafter. Because there are many more undergraduate institutions than Ph.D.-granting universities, the baccalaureate origins of Ph.D.s are less concentrated among larger institutions than are graduates by the Ph.D.-granting institutions we examined in Table 10.1. The top 175 baccalaureate origin institutions have approximately the same percentage of doctorates as the top forty doctoral institutions. The last column in this table gives the Herfindahl-Hirschman Index of the concentration of Ph.D.s by their bachelor's origins. The index falls slightly between 1970 and 1975 and then holds steady at 43–45, a level of concentration far below the comparable statistic for Ph.D.-granting institutions in Table 10.1.

In Table 10.3b the columns under "Percentage of U.S. baccalaureate–origin Ph.D.s by Carnegie Classification of Institutions" also show considerable stability in the proportion of Ph.D.s from the various categories. The same is true of data classified by Barron's Competitiveness of Institution in the next set of columns. In 1970, 2,519 graduates from the most selective schools earned S&E Ph.D.s. In 2000, 2,832 graduates from these schools earned Ph.D.s. This change roughly parallels the overall change in the number of U.S. baccalaureate–origin Ph.D.s. Finally, the last set of columns show stability in the share of Ph.D.s from universities with large amounts of federal research and development (R&D) money.

Beneath the stability in bachelor's origins by categories of higher educational institutions are changes in position of particular colleges and universities as source institutions for Ph.D. S&E graduates. These differences suggest an important role for college and university educational policies, including admission policies, in making some four-year institutions major sources for Ph.D. S&E doctorates. Among top-source undergraduate institutions, UC–Berkeley had 295 S&E Ph.D.s in 1970 and 279 in 2000—a modest drop. Cornell had 191 S&E Ph.D.s in 1970 and 262 in 2000—a marked increase. Harvard went from 223 S&E Ph.D.s in 1970 to 159 in 2000. MIT went from 280 in 1970 to 203 in 2000. The biggest decline in a Ph.D.-producing undergraduate institution was City College of the City University of New York (CUNY), which fell from the third-largest bachelor's

TABLE 10.3b Distribution of U.S. baccalaureate–origin Ph.D.s by three classifications of undergraduate institutional ratings, 1970–2000.

| Year | Percentage of U.S. Bacc.-origin Ph.D.s by Carnegie Classification of Institutions | | | | Percentage of U.S. Bacc.-origin Ph.D.s by Barron's Competitiveness of Institution | | | | Percentage of U.S. Bacc.-origin Ph.D.s by federal R&D spending received by Bacc. institution | | | |
	Research	Doctoral	Other institutions classified	Not classified	Most competitive	Highly competitive	Very competitive	Top 10 receivers	Top 25 receivers	Top 40 receivers
1970	56.47	10.93	32.10	0.49	16.91	24.06	7.57	11.81	21.36	29.04
1975	54.07	10.5	34.86	0.58	15.06	24.28	8.11	11.54	22.18	28.22
1980	55.41	11.15	33.04	0.41	16.2	24.29	9.18	10.99	21.97	28.28
1985	55.29	10.83	33.44	0.45	16.13	23.35	10.15	11.18	21.13	28.59
1990	55.63	11.07	32.78	0.52	15.14	24.09	9.95	11.63	20.86	28.6
1995	55.42	11.07	32.98	0.54	16.53	23.27	9.8	11.27	21.7	29.38
2000	56.10	10.48	32.99	0.43	18.06	24.4	9.58	10.27	22.59	30.45

Source: NSF, Survey of Earned Doctorates, various years, data available on WebCASPAR; Barron's, various years; Carnegie Commission on Higher Education, 1994.

origin institution of S&E Ph.D.s in 1970 with 245 doctoral graduates—10 percent more Ph.D.s than Harvard—to just 17 Ph.D.s in 2000. At the other end of the spectrum, Texas A&M University graduates earned 171 Ph.D.s in 2000, compared to 83 Ph.D.s in 1970, while the University of California, San Diego, undergraduates earned 152 Ph.D.s in 2000, compared to just 1 Ph.D. in 1970. Among smaller schools, Lehigh University went from 59 Ph.D.s in 1970 to 39 in 2000, while Harvey Mudd College went from 11 Ph.D.s in 1970 to 40 in 2000.

Variation among individual institutions notwithstanding, the principal conclusion from our analysis is that the bachelor's origins of Ph.D.s from U.S. undergraduate institutions barely changed over the period under study.[11] The increased share of foreign baccalaureates among S&E Ph.D.s, not reallocation of Ph.D. origins among those institutions, did reduce the shares of Ph.D.s originating from various categories of U.S. institutions.

The Flow of Students from Baccalaureate Sources to Ph.D.-Granting Destination Universities

The matrix that describes the quantitative relation between the number of persons graduating from particular bachelor's institutions and the number obtaining Ph.D.s from doctorate institutions is given by:

$$[A_{ij}]t$$

where A is an n by m matrix whose elements a_{ij} measure the number of persons with a bachelor's degree from baccalaureate institution i who obtain a Ph.D. at institution j. The n rows represent bachelor's source institutions while the m columns represent Ph.D.-granting destination institutions. The t refers to a given time period in which the Ph.D.s are granted. For example, one row in the matrix would measure the number of bachelor's graduates from the University of Southern California (USC) who earned an S&E Ph.D.; one column would refer to the University of Nebraska Ph.D. graduates; and the corresponding a_{ij} would show how many bachelor's graduates from USC earned Ph.D.s at Nebraska.

Using data from the Survey of Earned Doctorates and the NSF's Web-CASPAR data set, we estimated $[A_{ij}]t$ flow matrices for two periods of time, 1970–74 and 1995–99. Each of the matrices shows the number of undergraduates from different institutions that obtained Ph.D.s in the five-year window at specific doctorate-granting universities. We use a five-year window because many cells have only limited numbers of cases.[12] As an example of the elements in the matrix, the 1995–99 matrix showed that 17 University of Chicago undergraduates earned a science or engineering Ph.D. at MIT in that period, 94 Harvard undergraduates obtained a science or engineering Ph.D. at UC–Berkeley, and 10 Harvey Mudd graduates earned a Ph.D. at the University of California, San Diego.

Table 10.4 summarizes the linkages between particular types of undergraduate institutions and particular types of doctorate-granting universities from the flow matrices. It shows the percentage of all Ph.D.s who did their undergraduate training at a specified category of bachelor's institutions and obtained their Ph.D.s in S&E at universities in the specified categories in 1970–74 and 1995–99. Each row links different undergraduate origins to different Ph.D.-granting institutions. The first row for 1970–74, for instance, shows that 4.58 percent of Ph.D.s in S&E in that year were awarded to students from the ten largest bachelor's-origin institutions who obtained Ph.D.s at the ten largest doctorate-granting universities. The next column gives the comparable statistic for 1995–99. The next row gives the proportion of Ph.D.s granted to persons from the top twenty-five bachelor's-origin institutions and top twenty-five Ph.D.-producing universities. These two measures show the link between large undergraduate and large Ph.D. programs. Row 3 gives the proportion of Ph.D.s who did their undergraduate work at research universities and earned Ph.D.s from research universities. Row 4 gives the proportion of Ph.D.s who graduated from Barron's most selective undergraduate institutions and earned Ph.D.s from universities with ten or more top-rated Ph.D. programs. Row 5 gives the proportion of Ph.D.s who graduated from Barron's highly selective undergraduate

TABLE 10.4 Conditional probabilities of baccalaureates from specified undergraduate institutions' programs earning Ph.D.s at top Ph.D. institutions, 1970–74 and 1995–99.

Bacc. source institution	Ph.D.-granting destination institution	Percent of Bacc. source getting Ph.D.s in destination institution	
		1970–74	1995–99
Top 10 by size	Top 10 by size	4.58	3.36
Top 25 by size	Top 25 by size	13.59	11.14
Research institutions	Research institution	44.83	43.18
Most selective	10+ top Ph.D. programs	6.38	5.56
Highly selective	10+ top programs	6.79	5.38

Source: NSF, Survey of Earned Doctorates, various years. Data available on WebCASPAR.

institutions and received Ph.D.s from universities with ten or more top-rated Ph.D. programs.

The table shows a modest drop in the proportion of Ph.D.s awarded to persons coming from the specified pairings, which implies that some of the growth of Ph.D. production occurred outside this group of institutions. But the reason for the drop is not any weakening of the link between the relevant undergraduate institutions and Ph.D.-granting institutions. Rather, it is the rising proportion of Ph.D.s granted by less prestigious graduate schools that underlies the declines in the proportions in the table. Conditional on the number of Ph.D.s produced, universities with ten or more top-rated Ph.D. programs actually increased their share of students from the most selective and highly selective undergraduate institutions. In 1970–74 universities with ten or more top-rated Ph.D. programs drew 51 percent of their Ph.D.s with U.S.-bachelor's origins from the most selective and highly selective undergraduate institutions. In 1995–99, these universities drew 55 percent of their U.S. bachelor's–origin Ph.D.s from the most and highly selective undergraduate programs.

Table 10.5 presents a more detailed look at the link between undergraduate institutions and Ph.D.s in S&E at four leading doctorate universities: Harvard and MIT, for private universities, and UC–Berkeley and Michigan for public universities. The upper panel records the number of Ph.D.s granted by each of the institutions—disaggregated by those who came from Barron's most selective, highly selective, and very selective institutions—from other U.S. baccalaureate schools and from foreign undergraduate programs. These data show a mixed pattern of change in the undergraduate origins of S&E Ph.D. students. The number of S&E Ph.D.s from the selective undergraduate schools falls at the two private institutions but rises at the two public institutions.

The bottom panel of the table records the number of Ph.D.s granted to graduates from seven "most selective" undergraduate institutions. It shows that one important factor in the decline of Ph.D.s from the most selective undergraduate schools at Harvard and MIT is a sharp fall in the number of their own bachelor's graduates staying on for Ph.D.s. In 1970–74, 224 of Harvard's 1,575 Ph.D.s had been undergraduates at Harvard. In 1995–99, only 133 of Harvard's 1,591 Ph.D.s in S&E had received their undergraduate training at Harvard. At MIT the drop in the number of bachelor's graduates obtaining Ph.D.s at the school is even greater, from 370 to 191. But the fall in "own-Ph.D.s" is not limited to those institutions. Both UC–Berkeley and Michigan also saw a considerable decline in the number of

TABLE 10.5 Ph.Ds with specified bachelor's origins at leading doctorate universities, 1970–74 and 1995–99.

Undergraduate institutions	Doctorate universities							
	Harvard		MIT		Berkeley		Michigan	
	'70–'74	'95–'99	'70–'74	'95–'99	'70–'74	'95–'99	'70–'74	'95–'99
Most Selective*	724	596	766	581	473	647	318	347
Highly Selective*	308	289	321	263	803	735	690	491
Very Selective*	49	71	70	78	87	126	61	98
Other U.S.*	255	180	312	191	527	538	690	558
Foreign	232	399	480	499	754	670	368	775
Blank response	7	56	10	767	153	69	8	50
All PhDs	**1575**	**1591**	**1959**	**2379**	**2797**	**2785**	**2135**	**2319**
Harvard	224	133	55	55	80	94	44	16
MIT	47	51	370	181	64	80	24	28
Berkeley	55	54	40	65	462	292	28	38
Michigan	30	23	27	19	38	40	367	18
Stanford	26	41	22	36	41	56	12	20
Cornell	52	45	36	33	36	41	29	40
Princeton	42	52	34	31	25	55	17	19

Source: NSF, Survey of Earned Doctorates, various years. Data available on WebCASPAR.

Notes: Blank response means no baccalaureate institution was reported.
* By Barron's selectivity.

their own who stayed to get a Ph.D. Taking all Ph.D.-granting institutions in the flow matrix, we find that the percentage of Ph.D.s granted to persons with a baccalaureate from the same school fell from 14 percent to 10 percent, on average.

DESTINATIONS OF BACHELOR'S GRADUATES

We examine next the distribution of bachelor's degrees among doctorate-granting universities—the rows of the $[A_{ij}]$ matrix. For each of the bachelor's origin–Ph.D. destination groupings in Table 10.4, we calculated the conditional probability that graduates from a given type of bachelor's institution earned a Ph.D. at the specified types of universities. Each statistic in the table gives the percentage of S&E Ph.D.s from the bachelor's source category who earned their Ph.D. in the destination doctorate category. The probabilities that a bachelor's graduate earned his or her Ph.D. in the specified destination category are lower in all cases in 1995–99 than in 1970–74. This shows that graduates from large bachelor's-origin schools as well as research institutions and selective bachelor's programs have become less likely to obtain their Ph.D.s at major or high-quality Ph.D. universities over time. Graduates from the best bachelor's programs were more dispersed among Ph.D. programs in 1995–99 than they were in 1970–74.

Finally, we have tabulated the number of S&E Ph.D.s from the most selective four-year institutions using the Barron's categorization and divided the institutions into two groups: small liberal arts schools (such as Oberlin) and large research universities (such as Harvard). During the weak job market for bioscience Ph.D.s in the 1990s (Freeman et al. 2001), some academics and university administrators worried that undergraduates at leading research institutions were not pursuing bioscience careers because they observed first-hand the difficult economic conditions facing the postdocs and graduate students in science labs. By contrast, equally able and interested students at liberal arts colleges were reported to be pursuing bioscience careers because they did not have comparable first-hand information about economic prospects. The statistics in Table 10.6 are designed to assess this claim. The table records the number of S&E Ph.D.s, the number of foreign-born S&E Ph.D.s, and the number of bachelor's graduates five years earlier, at the two types of schools. In addition, it records the ratio of Ph.D.s in the given year to the number of bachelor's graduates five years earlier—a measure of the propensity of bachelor's from these undergraduate institutions to obtain doctorate degrees in the future. The statistics show a modest increase in the number of S&E Ph.D.s from the small liberal arts colleges relative to the larger research universities. The ratio of the number of Ph.D.s from the liberal arts colleges to the number of Ph.D.s from the large research universities in the period rises from 0.21 in 1970 to 0.25 in 2000. Similarly, the difference between the ratios of Ph.D.s to bachelor's graduates between research universities and the liberal colleges narrows over the period. Still, proportionately more bachelor's graduates from the research universities obtain Ph.D.s in science and engineering than do bachelor's graduates from the liberal arts colleges, and the ratio of Ph.D.s to bachelor's degrees remains higher for research institutions than for liberal arts colleges. While some liberal arts colleges that produce relatively many Ph.D.s, such as Reed and Oberlin, are included in Barron's "highly selective" rather than "most selective" group, expanding the institutions under comparison is unlikely to change the general picture of relatively little change.

TABLE 10.6 Science and engineering Ph.D.s from bachelor's institutions on Barron's list of most selective schools relative to bachelor's graduates, 1970–2000.

	20 small liberal arts schools				25 research institutions			
Year	Ph.D.s (a)	Foreign-Born Ph.D.s	Bacc.s granted 5 years earlier (b)	Ratio of (a) to (b	Ph.D.s (c)	Foreign-Born Ph.D.s	Bacc.s granted 5 years earlier (d)	Ratio of (c) to (d)
1970	442	13	7627	5.80	2077	142	22538	9.22
1975	422	24	8588	4.91	1814	105	24895	7.29
1980	468	26	10375	4.51	1808	141	29360	6.16
1985	440	32	11666	3.77	1770	129	31781	5.57
1990	452	30	12494	3.61	1779	178	33897	5.25
1995	495	38	12733	3.88	2128	249	35410	6.01
2000	572	55	12920	4.43	2260	341	35741	6.32

Source: NSF, Survey of Earned Doctorates, various years; National Center for Education Statistics, Higher Education General Information Survey (HEGIS); National Center for Education Statistics, Integrated Postsecondary Education Data System (IPEDS).

Notes:
Data available on WebCASPAR: http://caspar.nsf.gov/cgi-bin/WebIC.exe?template=nsf/srs/webcasp/start.wi
Bachelor's graduates reported for 1970 represent 1966 data.

CONCLUSION

This study has documented that the expansion of U.S. Ph.D. production in the 1970–2000 period involved changes in the demography of S&E Ph.D. recipients, the concentration of Ph.D. production, the birth origins of degree recipients, and the selectivity of both the sending and granting institutions, as follows:

1. The principal demographic development was that the share of U.S.-born white men fell among Ph.D. recipients, while the shares for U.S.-born women and minorities and the foreign born rose. In 1966 U.S.-born white males earned 71 percent of S&E Ph.D.s; in 2000, U.S.-born white males obtained 35 percent of S&E Ph.D.s. Larger shares of S&E Ph.D.s went to the foreign born, to U.S. women, and to U.S. minorities.

2. On the university Ph.D. production side, the proportion of S&E Ph.D.s coming from traditional leading doctorate institutions declined. This is not because these institutions took fewer students; the actual number they graduated remained approximately the same. Rather, it is because leading doctoral institutions did not expand production during the period so that growth of Ph.D.s occurred largely in less prestigious and smaller Ph.D. programs. In this respect, the educational sector responded differently than the response found in many other markets during an expansion, in which leading firms—those with the greatest brand recognition—play a major role in the expansion.

3. On the undergraduate source side, the proportion of Ph.D.s coming from various U.S.-source baccalaureate institutions has been relatively stable, with the major change being the reduced share of Ph.D.s coming from U.S. bachelor's institutions of all types due to the increase in foreign baccalaureate doctorates. In the most competitive undergraduate schools, there was a modest increase in share of Ph.D.s from liberal arts colleges relative to universities.

The observed changes reflect the interrelated decisions of students, Ph.D.-granting institutions, and undergraduate institutions, which are all influenced by government policies concerning the financing of education and R&D as well as by policies related to graduate stipends. These trends also reflect changes in the labor market for scientists and engineers and other high-level occupations that occurred during the time.

NOTES

1. The 71 percent figure in 1966 for males refers to "non-minority males," which we assume were almost exclusively white.

2. In 1973, 2,972 men did not answer the ethnicity question. Given the low proportions who answered minority, we assigned all who did not answer "U.S. citizen, white male" status. In 2000, 259 men did not answer the ethnicity question. We assigned all who did not answer to "U.S. citizen, white male" status. Calculations that simply ignore these respondents give results similar to those in the figures.

3. See www.usdoj.gov/atr/public/testimony/hhi.htm (last accessed April 16, 2007).

4. See www.carnegiefoundation.org/Classification/CIHE2000/background.htm (last accessed April 16, 2007).

5. In our data, there were 11,570 Ph.D.s in 1966 and 25,951 Ph.D.s in 2000, giving an increase of 14,381. The top forty Ph.D. producers granted 7,643 Ph.D.s in 1966 and 12,117 Ph.D.s in 2000. The "high-quality" institutions granted 5,579 Ph.D.s in 1966 and 8,289 Ph.D.s in 2000. The Carnegie research institutions granted 10,852 Ph.D.s in 1966 and 21,120 in 2000, so 29 percent of the increased number of degrees came from outside of that large group.

6. The forty largest Ph.D.-granting universities produce 50–60 percent of Ph.D.s, according to Table 10.1. Tabulating the data by other sized groups gives similar results to those in Table 10.2.

7. Black and Stephan (chapter 6 herein) report that the percentage of *temporary residents* graduating from a *top-ten Ph.D. program* declined during the period. There are two reasons why our result differs from theirs. First, their definition of a high-quality Ph.D. refers to specific programs in the top ten, while ours is much broader, referring to the quality of Ph.D. programs at the university and to large Ph.D.-producing institutions. Second, our analysis covers all of the foreign born, while their analysis focuses on those with temporary visas.

8. See http://caspar.nsf.gov (last accessed April 16, 2007).

9. The Carnegie Commission on Higher Education has periodically (1970, 1976, 1987, and 1994) classified institutions of higher education by the range of programs and/or degrees offered, enrollment size, and amount of federal funds received for research. We use the 1994 Carnegie classification to study the baccalaureate origins of scientists and engineers who received their doctorate from U.S. institutions. The changes to the 1994 classification were such that this analysis is not comparable to earlier Carnegie classifications

10. The Barron's *Profiles of American Colleges* guide differentiates colleges by their admission criteria. The "most competitive" colleges are determined by SAT scores, percentage of freshmen scoring who ranked in the upper fifth and the upper two-fifths of their high school graduating classes, percentage of applicants accepted for admission, and so on.

11. Analysis of the undergraduate origins of Ph.D.s by the NSF (1996) highlighted the proportion of foreign bachelor's graduates obtaining S&E degrees and

the concentration of Ph.D.s from twenty-five major bachelor's-granting institutions. Earlier analyses by the National Academy of Sciences examined the family background origins of Ph.D.s (NAS 1967).

12. We thank Teresa Grimes at the Quantum Research Corporation (QRC) and Keith Wilkinson at the NSF for suggesting that we use the five-year window and for creating matrices in useful forms in our initial tabulations.

THE CHANGING COMPETITIVENESS OF U.S. SCIENCE AND THE ORGANIZATION OF ACADEMIC SCIENCE

11

Global Research Competition Affects Measured U.S. Academic Output

DIANA M. HICKS

INTRODUCTION

Between 1992 and 1999, the number of papers published by U.S. academics fell by 9 percent as reported in the National Sciences Board's *Science and Engineering Indicators–2002* (SEI). This chapter seeks to understand why this occurred. A 9 percent decline in output could support the arguments of almost any constituency in U.S. academia. Advocates could report trends in particular fields over limited periods of time to support arguments about the deleterious effects of the emerging patent culture, the insidious effects of health insurers on medical research, the harm of decreasing federal support for engineering, the dangers of an aging university professoriate, and so on. This chapter approaches the question differently. It argues that to understand this decline properly we must take a step back and look at trends across the U.S. research enterprise. When we do, we see that the decline is so broadly based that any explanation particular to one field of research or even to universities as a whole must be inadequate. This chapter looks for a global phenomenon that explains both the broad pattern of decline and its surprising obscurity.

THE LARGER CONTEXT

Open publication in peer-reviewed research journals traditionally characterizes scientific communication, and counts of countries' papers in these journals are the basic indicators of national scientific output. Increases in published output are routine, expected, and taken as indicators of a healthy scientific research system. A declining publication count, however, would be

223

worrying, perhaps signaling weakness or decay in the research system, which in turn might threaten future economic growth in our science-driven, high-tech economy. SEI 2002 reported that U.S. academics published almost 8 percent fewer papers in 1999 than in 1995. Examined in the context of the past three decades, the decline is unprecedented.

Three decades of U.S. publication output as reported in SEI are displayed in Figure 11.1. Within each series, papers are counted only if they appear in journals indexed in the *Science Citation Index* (SCI) or *Social Sciences Citation Index* (SSCI)[1] in the first year of the series, taking into account journal administrative changes; that is, the journal set is fixed within series. Between series, the number of journals included in the counts increased resulting in the gaps between the series. The figure reveals that U.S. publication output has held steady within fixed journal sets and increased as the journal sets were updated. Only in the mid-1990s did decline set in.

The decline was broadly based, affecting more than just universities. Table 11.1 reports the percentage change in number of papers for each sector between 1992 and 1999. In aggregate, U.S. output dropped by 10 percent. Academic output dropped by 9 percent and accounted for 64 percent of the total change in output (because a large percentage of U.S. output comes

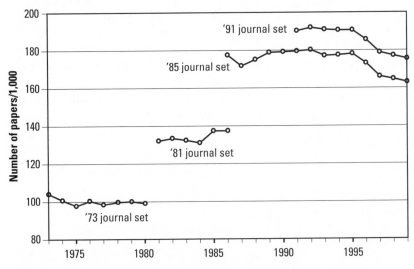

FIGURE 11.1 U.S.-authored papers fractionally counted in fixed journal sets, 1973–99.

Source: National Science Board, *Science and Engineering Indicators–2000.*

from universities). Note that federal government and corporate laboratories saw larger declines in output, and all sectors saw some decline.

The decline was broad in a second dimension; it affected most fields. For each scientific field, Table 11.2 reports the percentage change in number of U.S. papers between 1992 and 1999. Every field except earth and space sciences exhibited an absolute decline in output. The decline was most surprising in the life sciences, particularly biomedical research and clinical medicine, since these fields enjoyed increased government support during the 1990s. Granted, the declines in these fields were smaller in percentage terms than in other areas, but there were declines. Because the life sciences accounted for such a large share of papers, even their smaller absolute declines accounted for 41 percent of the total decline in U.S. output.

That it is not just university output in a few fields that was in decline in the late 1990s, but all U.S. scientific output across all fields, is a fundamental

TABLE 11.1 Percentage change in articles by sector, 1991–99.

Sector	Percent Change
Academia	−9
Federal Government	−17
Companies	−24
Nonprofits	−1
FFRDC	−1
Other	−13
Total	**−10**

TABLE 11.2 Percentage change in articles by field, 1992–99.

Field	Percent Change
Life Sciences	−7
Clinical medicine	−5
Biomedical research	−6
Biology	−22
Chemistry	−9
Physics	−9
Earth and space sciences	13
Engineering and technology	−26
Mathematics	−10
Social and behavioral sciences	−19
All fields/total	**−10**

point. Any cause must affect the entire U.S. research enterprise. Already doubt has been cast on particular constituencies presenting one or two of these figures to argue their case for more resources.

POSSIBLE EXPLANATIONS

The first possible explanation for the decline is that U.S. scientists are indeed publishing more and more articles in scientific journals as they always have, but SEI counts are somehow missing that and getting it wrong. After all, these publication counts are indicators, and indicators can go astray. Specifically, bibliographic databases index part of the scientific literature. No representation of the scientific literature is more complete than these databases, yet there are published scientific papers that are not indexed. Thus, publication indicators can go astray if the database is not a faithful representation of the scientific literature. Of particular relevance here, if the database grows more slowly than the literature, U.S. scientific output could appear to decline when measured in the database yet could still be expanding.

It is quite likely that the scientific literature grows differently from the literature databases because different mechanisms underlie the expansion of the literature and the expansion of databases. The literature expands as journals grow and split and new journals are founded to serve new specialist interests. This reflects the continuous expansion of research as scientists become more specialized and new specialties emerge. Perhaps most crucially, the literature grows as the result of decisions of innumerable highly motivated individuals and publishers looking for expansion. This growth may be slowed, however, if subscriptions are harder to come by—for example, if library budgets are static or journal subscription costs rise. Databases must expand or risk looking old fashioned and losing subscribers, yet expansion per se probably does not increase subscriptions, though it does increase costs. Also, databases are affected by changes in company policy, such as a decision to include more foreign literature or to add all health sciences journals (as happened in the SSCI in 1996). Database growth is controlled by skilled management who assess the costs and benefits. Thus, the literature is biased toward exuberant and uncontrolled growth, while databases are biased toward staying the same size or, if they must, growing slowly, steadily, and predictably. The SCI grows linearly at about 3 percent per year.

This disparity is not as worrying as it might seem at first because of what might be termed "the quality factor" in science. Simply put, not all literature that claims to be scientific is really scientific, and not all scientific literature is equally valuable. In the first category would be, for example, journals about astrology or homeopathy. In the second would be house journals, locally oriented journals, and often new journals, because researchers may be wary of submitting good work to new journals with an uncertain future and limited circulation. The boundary between the best scientific literature and the rest is subjective and shifts over time as approaches once considered obsessions of the fringe gain acceptance. However, from the policy perspective, quality counts. If U.S. output in *Science* and *Nature* declined, it would not be comforting to know that increased publishing in *Astrology Today* and *Vegetable Journal* more than made up for it.

To generalize this principle, in bibliometrics we rely on databases to draw a line somewhere and to incorporate the best scientific literature. We hope, and in general this is the case, that the database indexes the best literature and that at the bottom end of its quality spectrum questions may arise and coverage may change but that, overall, what is missed is much less important than what is indexed. This works because the impact of scientific research is not arrayed in a normal distribution (like height or intelligence). Measured by citations, impact follows a power law distribution, meaning that a very few papers earn very high citation counts and a large number earn no citations at all. Normal distributions are well described by a mean and standard deviation; power law distributions are not. Thus, database providers work not to cover a spread about a quality mean but, rather, to identify the top of the distribution. The top is very visible, given the nature of the distribution, and the nature of the distribution means that however far down the distribution databases draw their line, the literature excluded will be much less significant than the literature included.

The implications for assessing growth of U.S. scientific output are these. What we would like to count are papers in peer-reviewed, internationally oriented journals, because this is the yardstick by which a nation's science should be measured against other nations. The SEI fixed set of journals is relatively static in size when compared to the full scientific literature. Nevertheless, growth in number of U.S. papers is possible. Between the five-year periods ending 1994 and 1999, the number of papers in the SEI fixed set of journals grew by 5 percent. Although this growth rate is probably lower than would be obtained from a count of the full scientific literature, our indicators have not gone astray. If U.S. output in the world's top peer-reviewed, internationally oriented journals is in decline, that in itself may signify a problem. Our indicators may be more complex than we initially thought, but they raise policy-relevant questions.

ELECTRONIC PUBLISHING

Declining numbers of journal articles would not be cause for worry if authors were shifting to newer forms of publishing such as electronic publishing or new journals. The SCI indexes high-quality, regularly published electronic journals, so for this explanation to hold, something would have to be happening that defeats their policy of indexing the best literature. In addition, the tenure system at U.S. universities is rigid in its focus on journal quality, and this has not been said to be changing. So presumably, if this explanation holds, already tenured academics are responsible. Finally, for this explanation to hold, American scientists would have to be in the vanguard of shifts to electronic publication, depressing their journal article counts relative to those of foreign scientists, even though Web preprint archives are very helpful for geographically isolated scientists (Glanz 2001).

Although electronic publishing is much discussed, perspective is important. Kling and McKim (1999) point out that we must distinguish between pure electronic journals and hybrid paper-electronic (p-e) journals, or traditional journals whose contents are available to subscribers online. In the late 1990s there were remarkably few pure electronic journals, and they published few papers. For example, in 1997 the *Internet Journal of Science: Biological Chemistry* published fewer than 10 papers (and then disappeared) while the traditional *Journal of Biological Chemistry* published about 3,500 (and survived to put its full text online at: www.jbc.org). Kling and McKim (1999) suspect that

> reports of exponential growth of e-journals really mean exponential growth of p-e journals. This is not a minor matter, because the p-e journals bring their reputations, review practices that they established in the paper world, and some of their readership to their electronic versions. In contrast, new e-journals . . . face more daunting problems in establishing their legitimacy, and risk a higher failure rate. (892)

While questions remain about the legitimacy of publishing in pure e-journals, there is no evidence that the legitimacy of a paper journal declined when it became a hybrid paper-electronic journal. While the movement to hybrid paper-electronic journals reshapes scientific communication, it does not affect the bibliometric data. The indexing of traditional journals in the SCI is not affected by their adding an electronic version.

Perhaps the move to electronic publishing will bypass journals. The success of Paul Ginsparg's physics e-print server suggests that circulation of unrefereed papers might replace journal publication. Ginsparg and others have argued for the inevitability of this mechanism spreading throughout the sciences due to its cost advantages and evident superiority. However, Kling and McKim (2000) argue that it is "not just a matter of time" until all fields have an e-print server. They point out that communication practices differ between fields and suggest that the electronic communication projects that thrive tend to support and enhance preexisting practices. Physics had a thriving preprint exchange culture before the e-print server, for example. Molecular biologists never had broad circulation of preprints, limiting preprint circulation to their closest colleagues, and the few e-print servers in biology do not play a significant role in the biological communication system. In contrast, biologists increasingly depend on shared electronic databases that contain genomic data and associated information such as bibliographies, directories of suppliers, and contact information for researchers (Kling and McKim 2000). These databases are symbiotic with and enhance journal publication as journals increasingly require authors to submit information to these databases as a prerequisite to publication. Even e-print servers enhance scientific communication without supplanting traditional journals; 70 percent of the submissions to the e-print server in high energy and nuclear physics end up as journal publications.[2] Note that the U.S. decline would not be explained just by establishing that U.S. authors were moving to more informal means of publication. It would also be necessary to show that foreign authors were not doing the same thing. It is quite likely that informal publication is growing in the United States and abroad, and formal publication is growing abroad. Only U.S. formal publication is in decline.

Shifts in publication habits, if they are of any significance for the scientific literature, should leave their mark on that literature. Important work published outside SCI-indexed journals should be referenced by papers in established journals. This is the case for books in the social sciences. Although books are not indexed in the SSCI, they are so frequently cited in the journal literature that their role in the social sciences is clear. The same should hold for work published in electronic venues or in new journals not yet indexed in the SCI or SSCI. SEI 2002 reported that "an analysis of reference patterns in a sample of 986 papers published in 1990, 1995, and 1997 found few references to Internet URLs." This must cast a certain amount of doubt on the hypothesis that important work migrated to the Internet in the 1990s.

If a large amount of important research were being published electronically, it would change the pattern of referencing from patents to the literature. In particular, we would see a decreased share of references to journal articles and an increased share of references to other types of work. This was examined by CHI Research, Inc., which classified nonpatent references into "journals," "meetings," and "other." They calculated the share of references in each of the categories by year of referencing patent. They found that, over time, the share of patent references going to journals increased, which does not support the hypothesis.[3] Finally, one research institute's publication lists were analyzed, Woods Hole Oceanographic Institution. The institution's lists of published output in 1995, 1996, and 1997 were obtained from their library. The institution's output dropped in 1997 as counted from their own publication lists. Papers in both journals and in books and conference proceedings dropped in 1997. However, the lists contained no evidence of publication on the Internet.

Electronic publishing has been much discussed and therefore becomes an obvious hypothesis to explain the decline in U.S. publication output in the late 1990s as indexed in the *Science Citation Index*. However, the evidence does not substantiate the hypothesis. We suggest that if the effect of electronic publishing is too subtle to be found in the research reported here, then it is too subtle to cause a noticeable decline in U.S. publication output.

Rise in Academic Patenting

Patenting and licensing by universities has blossomed over the past ten years, as documented and discussed in chapters 3, 4, and 5 of this book (see also Hicks et al. 2001; AUTM 2000). Universities whose academics get involved in commercial activities must face the intractable problem of competition for academics' time. If patenting absorbs time academics would otherwise spend on research, the rise in patenting might be related to the fall in publication output. Logical though this sounds, the data seem to show no relationship between rates of patenting and rates of publishing for professors or universities.

Agrawal and Henderson (2001) studied the patenting and publishing behavior of professors in MIT's departments of mechanical and electrical engineering. They regressed patents and papers against each other with various lags and control variables such as length of time the professor had been active. They found no significant positive or negative relationship.[4] Stephan et al. (forthcoming) find a positive relationship between publishing and patenting, suggesting that they are complementary activities.

SEI 2002 reported the same thing at the aggregate level. Since universities differ in their promotion of and success in patenting, we might expect that growth rates in patenting will differ across universities. Universities with the highest rates of growth in patenting might be those with the greatest declines in publication output, if more patenting suppressed publication output. SEI 2002 reports that "there appears to be no significant difference in overall output of articles from universities that are major patentees and those that are not. The change in output of the former between the two three-year periods ending 1995 and 1999 was -5.4 percent compared with -4.6 percent for the latter" (5–41). Growth in patenting was uncorrelated with growth or decline in publishing, which suggests an absence of a connection between increased patenting and decreased publishing in universities.

DEMOGRAPHY

If older researchers are less productive, it might be that the aging U.S. scientific workforce is responsible for the decline in U.S. output. However, SEI 2002 reported that

> in the early 1970s, nearly half of all academic scientists and engineers were younger than age 40. Twenty years later, that figure had fallen to 28 percent, and by 1997, it had dropped to 25 percent. If age affects research productivity negatively, then this factor could provide a plausible explanation. However, the apparent decline in publications did not occur until after this demographic shift had been well under way during the previous two decades.

TRENDS IN U.S. RESEARCH FUNDING

U.S. research funding patterns have shifted over the past decade in ways that might explain growth and decline in publication output. Looking back at Tables 11.1 and 11.2 and generalizing, biology and the physical sciences seem to be suffering the most and the medical and environmental sciences prospering the most. This aligns with common knowledge of trends in federal research funding in the 1990s. Biomedical research has been a priority, while cutbacks at the Department of Defense would hit the physical sciences particularly hard.

Unfortunately, attempting analysis in more detail quickly becomes frustrating. Matching trends in funding to trends in paper output is complicated, first because the field classification schemes used for funding and papers differ somewhat, and second, because scientists receiving money that the government classifies as "chemistry" are not required to publish in a journal classified as "chemistry." Thus, trends in chemistry funding only roughly match trends in chemistry publishing.

SEI 2002 reported that for fields in which an approximate match could be made, the findings were inconclusive (5–41). For example, the fall in articles in biology and physical sciences coincided with a fall in federal spending (in real terms) in these two fields. However, increases in funding for physics coincided with a decline in articles. Matching funding and publication by sector is more straightforward, because institutions are classified the same way. However, there appears to be no correlation between these two variables. Basic and applied research expenditures have increased in universities and the federal government, but article output has declined in these sectors. However, funding increases in the nonprofit institutions and nonprofit Federally Funded Research and Development Centers (FFRDC) have coincided with increased article output in these sectors. A more precise match between the National Institutes of Health (NIH) publication output and intramural expenditures reveals that the trend of funding and publication growth diverged in the early 1990s, with publication growth flattening as funding continued to increase.

In conclusion, publishing in medical and environment-related areas has grown, suggesting that trends in domestic research funding shape trends in output, as we would expect. However, our expectations are confounded at the sector level, where trends in research expenditure and publication output are not aligned.

An International Perspective

If U.S. authors were winning less space in the top scientific journals, then foreign authors were winning more space. This is illustrated in Figure 11.2, which reports a growth index for publication output around the world between 1991 and 1999. Declining output was not confined to North

America. Publication counts declined more in Eastern Europe, which includes countries of the former USSR, and in sub-Saharan Africa. Everywhere else, publication output grew. Growth was quite striking in Latin America and most especially in the newly industrializing countries of East Asia: Hong Kong, Singapore, South Korea, and Taiwan. The consequences of growth in output abroad are most striking when European and American publication output are compared—Western Europe now publishes more than the United States.

That foreign scientific output grew so much is likely due to explicit national policies. Foreign science systems have strengthened in recent years as governments around the world recognized the need to build knowledge-based economies and, as part of this, have increased research funding, strengthened graduate programs, and started to evaluate their scientific output more stringently. Most fundamentally, resources have been dedicated to science and technology, especially in East Asia. SEI 2000 reported:

> Several Asian countries—most notably South Korea and China—
> were particularly aggressive in expanding their support for
> R&D and S&T-based development. (2–47)[5]
> In Latin America and the Pacific region, other non-OECD
> (Organisation for Economic Co-operation and Development)

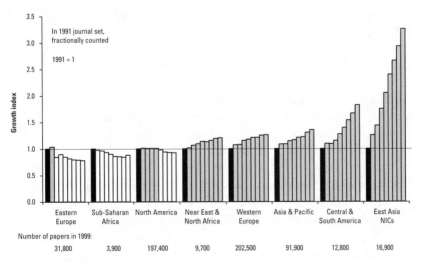

FIGURE 11.2 Growth in number of papers worldwide, 1991–99.

Source: National Science Board, *Science and Engineering Indicators–2000.*

countries also attempted to increase R&D investments substantially. They still invest less than European countries in R&D. However, they also have substantial S&T-related government expenditures not captured in R&D statistics, especially expenditures on training and infrastructure. (2–47)

There was a worldwide slowing in R&D spending in large and small countries in the early 1990s. In fact, inflation-adjusted R&D spending fell for three consecutive years (1992, 1993, and 1994) in the United States, Japan, Germany, and Italy. R&D spending has since recovered in these countries but has remained stagnant in France and the United Kingdom. Most of the recent R&D growth results from rebounding industrial nondefense spending. (2–4)

The most notable trend among OECD countries was the relative decline in government R&D funding. (2–4)

R&D spending in the Russian Federation remained considerably below levels in place prior to the introduction of a market economy. (2–4)

Thus, it is no surprise that in the late 1990s East Asian countries topped the list of fast-growing publishers, and Latin American countries were increasing their publication output. Since government research spending should bear the most direct relation to publishing, we would not expect a rise in Western countries whose R&D growth traces to increased industrial R&D. The Russian Federation is a clear case of declining research resources and output.

Graduate programs have been strengthened as well. In 1998 the National Science Foundation held a workshop on graduate education reforms in Europe, Asia, and the Americas and on international mobility of scientists and engineers. Johnson and Coward (2000) in their overview of the discussions pointed out the following:

France undertook a reform of doctoral studies in 1988 in an effort to double the number and improve the quality of S&E doctoral degrees awarded within eight years. By 1996 they had achieved a 75 percent increase.

In Japan, doctoral reforms in 1989 called for expanding and strengthening graduate schools and for establishing a new type of university exclusively for graduate study. By 1994 more Japanese engineers earned doctoral degrees in university laboratories than within industry, which had been the dominant route.

China has invested heavily in graduate education with the result that in 1997 Chinese students earned more than twice as many S&E doctorates within Chinese universities as did Chinese students within U.S. universities.

South Korean universities awarded almost 2,200 S&E doctoral degrees in 1997, up from 945 in 1990.

S&E doctoral degrees earned by Asian students within U.S. universities peaked in 1996 and declined in 1997. S&E doctorates earned in Asia grew by 12 percent per year on average between 1993 and 1997.

Within Latin America, Brazil greatly expanded the scale of its graduate programs in the 1980s, and Mexico, Chile, and Argentina have done so more recently. These countries are motivated by a desire to have more of their university faculty trained at the doctoral level.

Note that we know much less about the science of countries that have hitherto not made much impression in the scientific world. Many may be greatly increasing the resources they devote to research. For example, Egypt built three glass and concrete pyramids to house the Mubarak City for Scientific Research and Technological Application (MUSCAT). Institutes on everything from lasers to desert research are slated for launch (Frank 2001). If obstacles can be overcome, and expatriates attracted, ventures like this could change the scientific landscape in coming decades. We may already be seeing the first evidence of their impact on publication figures.

It may also be important to acknowledge that research funding may not fully explain the strengthening of foreign science systems. Increasingly, foreign governments tie research funding to output evaluation. When foreign governments go down this route, journals indexed in the SCI or SSCI often become the gold standard. Governments want scientists to publish in these journals; they make that clear. And scientists respond by focusing their efforts on publishing in indexed journals. This makes sense for governments that want their scientists to be working at an international standard and not lurking in the local literature, which may privately be acknowledged to be less than scholarly.

The United Kingdom may have been one of the first countries to implement such a system. Most recently, China has been implementing such evaluations, *Science* has reported:

> As scientific activity recovered after the Cultural Revolution, much of the funding flowed from the top down, split up more or less by seniority. But competitively awarded grants now predominate, with emphasis on a good track record. . . . Although base pay still depends on rank, explains NIGP (Nanjing Institute of Geology and Paleontology) director Sha Jingeng, publications, prizes won, and research grants awarded also play a role. For each paper that appears in *Science* or *Nature*, for example, NIGP pays a researcher about $600 (Normile 2001).

Although the shake-up in China is broad based, institutes devise their own implementation plans.

> For instance, officials at a key national lab of the Institute of Zoology in Beijing began annual evaluations of staff based on the number of projects worked on, research grants obtained, international conferences and collaborations, students being advised, and papers published in Chinese and foreign journals traced by the *Science Citation Index* (SCI). The last . . . is seen as an external measure of quality. "This system can help us judge a researcher's work more objectively and fairly," says Li Dianmo, the lab's director (Yimin 2001).

Although U.S. scientists benefit from publishing in top journals, Chinese scientists who land among the top half of their colleagues ranked using measures such as those just described can earn three to four times the salaries of their coworkers. In the United Kingdom, academic departments get more money if expert committees award departments high scores based in part on limited bibliographies of their best published output. Because Australia uses a formula to distribute university funding, Australian universities can value in dollars each article faculty members publish in an indexed journal. Australian ISI-indexed output rose in response to the introduction of this formula (Butler 2002). Scientists are enticed by explicit incentives to target SCI-indexed journals for article submissions. Highly motivated foreigners targeting SCI journals will make life more difficult for U.S. researchers. U.S. scientists have always been highly competitive, so there is little scope for the kind of performance gains foreign systems appear to be achieving.

The hypothesis that an international perspective is needed to understand the decline in U.S. output is supported by quantitative data on research resources such as funding, researchers, and students. Domestic perspectives on research resources are given in Figure 11.3, which examines trends in total U.S. R&D expenditures, numbers of scientists and engineers, and numbers of graduate students and doctoral degrees awarded.[6] In Figure 11.4, the international perspective, the U.S. resources are expressed as a share of a group of countries.

In the domestic perspective shown in Figure 11.3, there is a general pattern of growth with perhaps some leveling and a bit of decline, but nothing especially worrying. Thus, long periods of growth are interrupted in the early 1990s (except in doctoral degrees awarded, which continued to grow). There was a slight decline in R&D expenditures, a leveling of growth in scientists and engineers, and a drop-off in graduate enrollment. Note that development expenditures and personnel are included in these figures, and as military expenditure declined substantially during this period, an unknown proportion of the drop must be due to military development, which would not affect the trend in paper output much.

The most worrying decline is the drop in number of graduate students beginning in 1993, which followed four decades of increase. Historically, growth in graduate enrollment has generally echoed shifting patterns of federal R&D support, with an influx of foreign students complicating matters. Some of this decline may trace to the favorable U.S. job market after

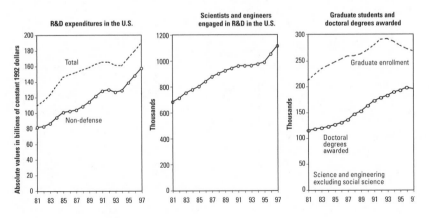

FIGURE 11.3 U.S. R&D metrics, 1981–97.

Source: National Science Board, *Science and Engineering Indicators–2000,* appendix tables 2-63, 2-64, 4-21, 4-28, 4-29.

1992, especially since the decline is most concentrated in engineering and math and computer science. Some of the decline undoubtedly relates to increased opportunities non-U.S. students have to study outside the United States. The decline in graduate students may well be connected to the decline in university output, though the decline in graduate enrollment seems to affect the biological sciences as well as the physical sciences, which does not echo the pattern in paper output (National Science Board 2000, 4–20).

In the domestic perspective, there are hints of problems, but things really do not look too bad. In Figure 11.4, the data on U.S. research resources are expressed as a percentage of a group of countries that includes the United States.[7] This international perspective is truly worrying. The U.S. share of G7 R&D expenditure declined sharply from 1987 to 1990; the U.S. share of G7 scientists and engineers declined sharply from 1991 to 1993; and the share of U.S.-Asian doctoral degrees began declining sharply in 1993. The international perspective reveals patterns very similar to the drop in U.S. share of publications—a long-term, gentle decline in share is rudely interrupted by a slide down a much steeper slope. The declines in resources begin a couple of years before the declines in U.S. output, which makes perfect sense. Share of R&D expenditure and scientists and engineers has since turned up, offering hope that the number of U.S. publications might rise again.

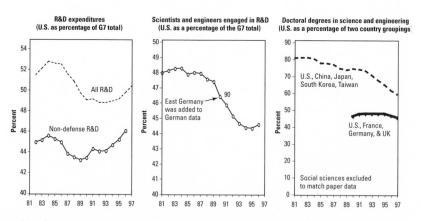

FIGURE 11.4 R&D metrics: United States as a percentage of international values, 1981–97.

Source: National Science Board, *Science and Engineering Indicators–2000*, appendix tables 2-63, 2-64, 4-21, 4-28, 4-29; figure 3-18 (U.S. data interpolated in even years).

DISCUSSION

Since counts have begun, the U.S. share of world publication output has declined while the number of U.S.-authored papers has increased or held steady, depending on the counting method. The late 1990s saw a new pattern, an absolute decline in number of U.S. papers in the international, peer-reviewed literature.

The late 1990s were notable for the rise of the Internet and for increased patenting by U.S. universities. Both offer plausible explanations for the decline. Perhaps publishing on the Internet is replacing traditional journal publication? However, there appears to be no evidence that in the 1990s U.S. scientific output moved to electronic venues. In any case, if U.S. output in *Science, Nature, PNAS, Cell,* and other top journals declined, would electronic preprints be an adequate substitute? Possibly, then, patenting is to blame. Do university researchers patent more by publishing less? But available data at the individual and institutional levels does not support this idea.

As it happens, the rise of the Internet and of public-sector patenting, though dramatic, are perhaps not yet as powerful an influence on scientific work as money. Recent trends in federal funding have been suggestively similar to trends in publishing in that biomedical sciences have fared much better than natural sciences in both publishing and securing federal funding. However, this is not conclusive, as federal funds are not the only source of research funding in the U.S. and the decline is very broad in scope. For example, university expenditure on research has risen annually for a decade or more, belying any close correlation between publication and funding trends. Trends in U.S. research funding provide one piece of the puzzle but not the whole story.

To understand fully the decline in U.S. publication output, an international perspective is required. Publication output is increasing across a broad range of foreign countries—newly industrializing East Asian countries, Latin America, Asian and Pacific countries, and Europe. North America, sub-Saharan Africa, and Eastern Europe exhibit declines. In the 1990s, growth in scientific publishing by other countries accelerated, forcing down the U.S. share of the top journals. U.S. and foreign authors compete in an almost zero-sum game for places in journals indexed in the SCI or SSCI. If foreigners are aggressive enough in increasing their share of publications in top journals, and if U.S. authors maintain their status quo, U.S. authors could lose share so quickly that growth in the database will not compensate for their lost share, and the number of U.S. papers will decline. This would produce a decline in U.S. publication output almost irrespective of domestic policy or sector-specific factors, and we do see a very broad decline.

International data on R&D expenditures, scientists and engineers, and doctoral degrees, though imperfect, tends to echo the publication data. Or rather, the pattern of gradual decline in the U.S. share interrupted by an abrupt shift to a much steeper decline is seen in the U.S. share of money, people, and students and is echoed with a convincing lag of a few years in the publication data. Moreover, the quantitative evidence is compatible with our admittedly imperfect knowledge of trends in foreign science policy. During the past decade, other countries have focused resources on research, expanded graduate programs, and sought to restructure their public-sector research institutions as part of a drive to create knowledge-driven economies.

The importance of international forces in explaining the decline in U.S. publication output accounts for two somewhat unusual aspects of the decline, namely its breadth and its invisibility. Reasoning from knowledge of trends in domestic funding, we would expect federally and commercially funded physics to be in steep decline and federally and commercially funded biomedicine to be growing buoyantly. However, federal physics laboratories are not the only ones whose output has declined; almost everybody is suffering, including pharmaceutical companies. Even NIH laboratories do not show the expected strong growth in output in recent years. Among domestic factors, only demographics could act with such breadth, but demographic trends would not exhibit the abrupt shift we see in the publication data.

The second curious aspect of the decline is that there has not already been an outcry over a crisis that has been brewing for half a decade. Admittedly limited experience suggests that the idea of an overall decline in U.S. output strikes scientists as strange. It makes sense if they think about it, but is not something that would have occurred to them. Perhaps this is because the decline manifests itself to scientists as an unexpected rejection of a paper or two. Highly competitive and competent researchers would probably blame only themselves for this. Why would they think to blame a decade of concerted effort by foreign governments to improve their scientific infrastructures?

Conclusion

We seem to be entering a new era in science policy. The United States has long accepted that its share of world scientific output will decline as scientific communities in other countries strengthen. This process seems to have accelerated sharply, as other governments have become convinced that their economic futures lie with knowledge-based economies in which research plays a central role. Foreign scientific communities have become much more competitive at the same time as the U.S. federal government's attention is increasingly focused on sciences closely linked to medical care.

This chapter's data would fit neatly into the Council on Competitiveness's 2001 analysis of America's relative international standing in capacity for innovation. Drawing on Porter and Stern's Innovation Index, the council argued that in the 1990s the economic landscape began to change and a number of advanced nations increased their capacity for innovation and began to converge on the United States. In addition, new groups of innovative countries and fast followers emerged. Drawing on a range of indicators, they argue that the United States' position is slipping (Porter and van Opstal 2001).

As foreign scientists note their newfound prowess (Kocher and Sutter 2001), American scientists face the challenge of the steady state. Ziman (1994) first pointed out that scientific institutions are predicated on continued expansion. For example, each faculty member trains many students who then require employment. As research advances, new specialties are created that require more resources over and above those needed to keep up with inflation as ever-more-complex equipment is needed to solve the ever-more-complex problems left behind as simpler problems are solved. Therefore, adjustment to a "steady state" or a decline in resources is extremely difficult. Although Ziman was discussing resource limitations rather than output limitations, we can predict that output declines will make life much more difficult for already stressed U.S. scientists.

Signs of strain have appeared, as Donald Kennedy's (2001) reflections on his first year as editor of *Science* attest. After listing morally dubious practices encountered in his first year, he writes:

> Since we all know that such things happen, why call attention to them now?
>
> The reason is that their frequency appears to have increased. I think I know why. The universe is larger, and in the "hot" fields like molecular biology the competition—for funds, for appointment, for tenure, and for prizes—is more intense. And the advantages that accrue to publication in a prestigious journal are correspondingly large. In some countries, governments allocate prize money and promotion directly to researchers who publish in *Science*. In the United States and Europe the rewards are more subtle, but nonetheless real.

Under increasingly competitive circumstances outside of their control, can U.S. scientists and scientific institutions maintain the virtues so admired by the public and foreign scientific communities—fairness, openness, intellectual honesty, and rigor—virtues so necessary for swift and sure advance at the frontier of knowledge?

The institutions of modern science have in many ways been a gift from the United States to the rest of the world. The U.S. has demonstrated that the best-quality scientific research is fostered when funding is awarded competitively, when plentiful, rigorously trained Ph.D. students and postdocs are available cheaply, when substantial amounts of money are spent, when modern equipment is used, and when transfer of research to technological application is encouraged. In many ways, other countries have sought over the past decade to incorporate more of these elements into their systems. Furthermore, the U.S. has probably trained, or at some point employed, many of the scientists now doing so well back home. As a result, American universities no longer stand alone at the scientific frontier.

Notes

1. The *Science Citation Index* (SCI) and the *Social Sciences Citation Index* (SSCI) are products of Thomson Scientific (formerly known as Thomson ISI) and are the standard source for bibliometric citation analysis.

2. Seventy percent of 1999 submissions to physics e-print server in SLAC's core fields of high-energy physics and nuclear physics end up in journals. Unpublished data provided by Heath B. O'Connell, SLAC, Stanford, Calif.

3. Personal communication from Tony Breitzman, CHI Research, Inc., a scientific research firm that specialized in patent citation analysis.

4. They did, however, find a significant relationship between citations to papers and patenting. Professors whose papers had more impact produced more patents. This is consistent with Hicks et al. (2000, 310–20) who found that patents tend to preferentially cite more highly cited papers. In both cases we see a relationship between high-impact science and patenting.

5. Reported in Johnson (1993).

6. Note that social sciences have been excluded from these figures.

7. It was not possible to use a consistent group of countries across the graphs.

12

The Workforce for Biomedical Research—Who Will Do the Work?

SUSAN A. GERBI AND HOWARD GARRISON

CHANGES IN BIOMEDICAL RESEARCH

These are exciting times for biomedical research, as major advances in technology (e.g., sequencing, PCR, transgenic animals) and a wealth of data (e.g., genome sequences) allow rapid advances. In the second half of the twentieth century, the era of descriptive biology evolved into the modern age of reductionism, where fundamental principles and underlying mechanisms began to be elucidated at the molecular level. While much still remains to be learned at this basic level, the field has matured sufficiently to allow explanations of complex and interactive systems. This change in the direction of research requires adjustments in how we train the next generation of scientists. They will need breadth as well as depth in their education. Moreover, they will have to function as team players on "big science" projects in addition to individual investigator-driven research. Another change is that the gulf between basic research on fundamental biological mechanisms and translation of these findings to clinical applications has narrowed considerably.

There has been a large increase in the number of basic scientists employed as faculty members in clinical departments (Yamagata 1999; Fang and Meyer 2003). Additionally, significant interactions with biotech and pharmaceutical companies are not uncommon for basic biomedical scientists, and universities are eager to encourage faculty efforts in translational research. Some believe that this can lead to patents that could be financially rewarding for the university.

The exciting scientific prospects have captured the imagination and support of Congress, resulting in a doubling of the budget of the National

Institutes of Health (NIH) during the five-year period from 1998 to 2003, from $13.6 billion to $27.3 billion with almost half of these funds going to extramural research project grants (NIH Web site). The increasing complexity and specialization of biomedical science has led to increased collaborations on research projects. The trend toward "big science" has been facilitated by the increased funding, which made larger lab groups possible. The advantage of large labs in the competition among scientists has also led to growth in their size and complexity.

COMPOSITION OF THE WORKFORCE FOR BIOMEDICAL RESEARCH

Given the situation just described, what is the source and composition of the workforce that carries out laboratory research in academic settings? Academic research laboratories are structured as a pyramid (Goldman and Massy 2001). The faculty principal investigator (PI) is at the pinnacle, while the base is composed primarily of graduate students, leading upward to postdoctorals. The students and postdoctorals generally aspire in their career goals to move "up" the pyramid and become a faculty PI. However, it has been noted that a reasonable chance for success in such career aspirations will only be possible when the enterprise is expanding and there is growth in the number of new faculty positions (Goldman and Massy 2001). Although some will seek careers in other sectors (a point neglected in the analysis by Goldman and Massy), nonetheless, at a steady state not all of those in the pyramid will obtain a faculty position. While it can be debated when we might reach this steady state, it is already apparent that there is increasing unhappiness among postdocs who have unfulfilled career expectations. Concerned about this situation, some observers have recommended that the rate of entry into the pyramid be slowed down (National Research Council 1998). We have countered that this is not a reasonable solution, as it will simply result in more postdocs who received their Ph.D.s in other countries being imported (Gerbi et al. 2001). The challenge before us is to determine the optimal composition of research personnel in the academic laboratory. This challenge becomes even more pressing as the width of the pyramid increases with the expansion of "big science."

In the following sections, we examine various categories of laboratory personnel, listed in order of increasing training, and their role in fulfilling the academic laboratory's research mission.

Undergraduates

Undergraduates are wonderfully enthusiastic but need much training. Moreover, their short stays (typically one year) in the lab results in very little productivity return for the investment in training. Their cost to the PI can be great in terms of time but minimal in terms of dollars.

Master's Students

Master's students are not commonly found at most of the top research universities, as students typically enter into doctoral programs directly after their undergraduate training, bypassing a master's degree. Master's students are only slightly more advanced than undergraduates and only remain in the lab for one to two years for their thesis research. This relatively short training period results in only modest productivity relative to the training period.

Research Assistants

Research assistants are usually employed directly after an undergraduate or master's degree is conferred. Often these individuals are very bright and motivated, using laboratory employment as a short (one or two year) hiatus prior to enrolling in a Ph.D. program. In such situations, their length of stay in the lab is not very long. In other cases, people opt for this as the career goal, and with increasing years of experience they can become highly valued members of the lab. This source of labor is used to a greater extent in industry than in academia. Typically, the salary for these positions has been comparable to that of a postdoc.

Predoctoral Students

Predoctoral students are an extremely important part of the research lab workforce. They are very bright and highly motivated, willing to work long hours. Moreover, since predoctoral/graduate students have close ties (through initial coursework) with their peers in other labs at the school, they help the flow of techniques and ideas from one lab to the next.

Although many predoctoral students enter graduate school with only a little prior experience in research, their stay of several years in the lab where they carry out their research means that the time invested in training them by the PI pays off in their senior years in this position. The time to Ph.D. degree in the biological sciences after matriculation is approximately seven years at the national level (Figure 12.1), though several programs report 5.5

years as closer to their average (Marincola and Solomon 1998a). Nonetheless, there has been concern that the length of training is unduly long, and efforts are being made to shorten the period. A new program opened a few years ago at the Cold Spring Harbor Laboratory (New York) with a stated goal of graduating their Ph.D. students in four years.

Predoctoral students are usually paid through institutional funds (fellowships, teaching assistantships, and/or through NIH training grants) in the early years of their training, and costs are only born by the PI's research grants in the later years of training. This cost sharing is financially attractive to the PI. In addition, institutional funds may be available as bridge support if there is an interruption in grant support. This gives stability and continuity to the research program.

From 1987 through 1996, there was a substantial increase in the number of Ph.D.s being awarded in the biomedical sciences; some of the increase can be accounted for by an influx of foreign students into biomedical Ph.D. programs (Garrison et al. 2003) (Figure 12.2). The escalation in the number of biomedical Ph.D.s being produced was cause for concern, as it was feared that their numbers would outpace the career opportunities. One study recommended that further increases in graduate student production

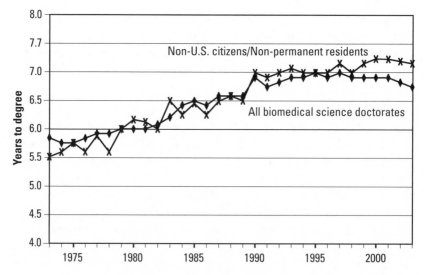

FIGURE 12.1 Time required to earn U.S. biomedical science Ph.D., 1973–2003.
Source: NSF, Survey of Earned Doctorates, 2003.

should be halted and no new graduate programs should be inaugurated (National Research Council 1998). In many European and Asian nations, the number of doctorates awarded in the natural sciences and in engineering has risen, eliminating or reducing the United States' lead (Figure 12.3).

Surprisingly, despite the excitement in biomedical research, the increase in biomedical Ph.D. production has now reached a plateau in the United States with about 5,000 Ph.D.s being awarded annually since 1996 (Garrison et al. 2003) (Figure 12.2). Is this due, in part, to the gloomy picture that was painted about the job market in the United States for biomedical Ph.D.s? A decreasing proportion of graduate students come from top undergraduate programs in the United States, as Freeman, Jin, and Shen document in chapter 10 of this book. Does this trend reflect the fact that undergraduates are exposed to intense competition for jobs and for grants? A new program of Howard Hughes Medical Institute Professors attempts to counter the downward trend by supporting professors with innovative ideas to nurture and inspire talented undergraduates to obtain a Ph.D. in the biomedical sciences. With an expanding biomedical research enterprise, the slowing in biomedical Ph.D. production implies that other types of laboratory personnel must be sought to carry out the work at the research bench.

FIGURE 12.2 U.S. Ph.D.s awarded in biomedical sciences, 1973–2003.

Source: NSF, Survey of Earned Doctorates, 2003.

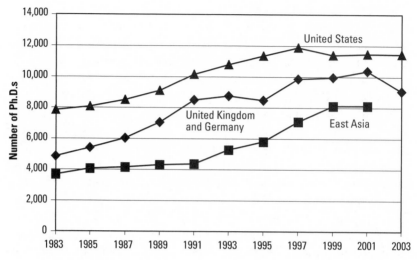

FIGURE 12.3 Physical and biological science doctorates awarded in Europe and Asia, 1983–2003.

Source: National Science Board, *Science and Engineering Indicators 2006.*

Postdoctorals

Postdoctorals, like graduate students, are bright and highly motivated. Postdoctorals are highly valued because they already have significant research experience and training. As an outcome of their predoctoral research, they often bring new ideas and methods to the postdoctoral laboratory. Postdoctoral costs to the PI's research grant are roughly comparable to those of a research assistant or a predoctoral student (with tuition). Moreover, many postdocs who are U.S. citizens succeed in getting their own fellowships and thus are not charged to the PI's research grants.

The postdoctoral experience is meant to be a time for further advanced training, increasing the breadth of research expertise. It is generally considered a prerequisite to a faculty position in academia. The fraction of biomedical Ph.D.s who choose to engage in postdoctoral training grew from approximately 30 percent in the 1970s to approximately 60 percent in the 1990s (Garrison et al. 2003) (Figure 12.4). There was a concomitant increase in the length of the postdoctoral period. About 20 percent were still postdocs five to six years after receiving their Ph.D.s in contrast to a negligible percentage in this category thirty years ago. The actual percentage may be higher, since the position title is sometimes changed for people with more than five years of experience.

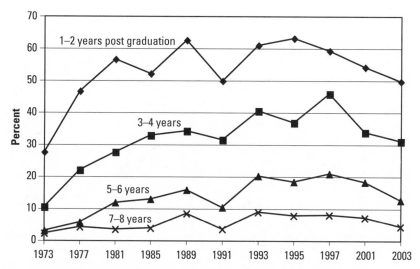

FIGURE 12.4 U.S. biomedical science Ph.D.s in postdoctoral appointments, 1973–2003.

Source: NSF, Survey of Doctorate Recipients, 2003.

Note: Scale change reflects change in survey frequency after 1989.

Some believe that the increased length of postdoctoral study reflects a holding pattern while postdocs seek a permanent job. Is the publicity that has been given to the plight of the postdoc with frustrated career objectives a reason that there has recently been a reversal of trends and a downturn in the number of U.S.-trained biomedical Ph.D.s going on to postdoctoral positions? In 1973, only 27.4 percent of the U.S. biomedical Ph.D.s took postdoctoral positions; this grew to a high of 63.3 percent in 1995 but had declined to 49.8 percent by 2003 (Garrison et al. 2003) (Figure 12.4). Similarly, there has been a recent decline in the length of stay as a postdoc. The percentage of biomedical Ph.D.s in postdoctoral positions three to four years after receiving their degree reached a high point of 45.9 percent in 1997 but dropped to 31.0 percent by 2003 (Garrison et al. 2003) (Figure 12.4). A growing number of graduate students take jobs in industry rather than academia, and many realize that postdoctoral experience is not always required for the former. The shortfall of U.S.-trained postdoctorals is made up by an influx of biomedical scientists holding Ph.D.s from universities in other countries (Garrison et al. 2003) (Figure 12.5). This population is not tracked by current surveys, and it would be interesting to know more

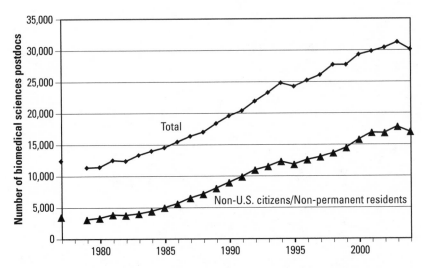

FIGURE 12.5 U.S. biomedical sciences postdocs at doctorate-granting U.S. institutions, 1972–2004.

Source: NSF, Survey of Graduate Students and Postdoctorates in Science and Engineering, 2004.

about their career outcomes. We do know, however, that 71 percent of the foreign students who obtained their Ph.D.s in the United States remained here in 2001 (Finn 2003).

Research Scientists

Research scientists are experienced individuals who are employees of research grants. Often they retain a postdoctoral title even though they are no longer in training. It has been recommended that their salaries be increased to reflect their experience and that their position title be changed from "postdoctoral" to "research scientist" (Marincola and Solomon 1998a and 1998b; Gerbi et al. 2001). They bring technical and intellectual sophistication and maturity to the research project and stability to counter the flux in the other sources of personnel. Usually only those who have proven their prowess in research as postdocs are chosen to stay on in these senior research positions, and they move the research project forward efficiently and skillfully. From the point of view of the PI, such personnel are invaluable. From the perspective of the research scientist, this position offers the enjoyment and intellectual challenges of bench research without the distractions of other duties such as teaching or administration. Moreover, research scientists generally do not have to write grants, as this is done by the PI. However, in addition to the relatively low salary, there is a lack of job security. Holding down this job requires continuous grant funding. In addition, when the PI retires, most schools have no commitment to the research personnel in his or her laboratory. By this career stage, the research scientist is highly focused and specialized with regard to research, and it can be difficult to relocate to other positions, especially as the employer may prefer entry-level personnel with lower salaries.

Sectors of Employment for Biomedical Ph.D.s

When most students who entered the workforce pyramid had a faculty position in academia as their career goal, postdoctoral training was the norm. However, as other job opportunities have become available and more popular as options, students are beginning to question whether they need to take postdoctoral positions. What are the job sectors where biomedical Ph.D.s are employed, and how have they changed over time?

Academia

In 2003, 52 percent of biomedical Ph.D.s were employed in academia. While the absolute number in academia has grown a bit since 1985, the percentage has decreased due to opportunities increasing more rapidly in other sectors (especially in industry). Moreover, the composition of the academic pool has changed (Garrison et al. 2003) (Figure 12.6). There has been a decline in the number holding tenured faculty positions from 25,104 in 1985 to 23,088 in 2003 (Figure 12.7). Although within the basic sciences departments of U.S. medical schools, half of new faculty hires are now non-tenure-track positions at the rank of assistant professor or above (Liu and Mallon 2004), when looking at universities and colleges overall, the number of biomedical scientists in non-tenure-track faculty positions has grown modestly from 7,795 to 9,578 (Garrison et al. 2003). It is noteworthy that there has been a marked rise in those holding "other" academic positions from 9,704 in 1985 to 18,700 in 2003 (Garrison et al. 2003) (Figure 12.7). Although the exact type of job held by people in this category is not reported in national surveys, it likely includes the senior postdoctorals who are given a different title (e.g., research scientist; research-track assistant professor, etc).

It is of great concern that the number of PIs who are thirty-five years old or younger competing for R01, R29, and R37 research grants has dropped from 8.4 percent in 1991 to 1.9 percent in 2000 (Moore 2002). This is especially grave as some of the most original and creative work is often

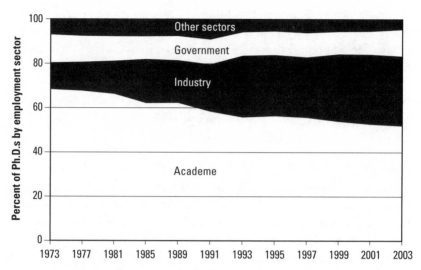

FIGURE 12.6 U.S. biomedical science Ph.D.s employed in industry, 1973–2003.

Source: NSF, Survey of Doctorate Recipients, 2003.

Note: Scale change reflects change in survey frequency after 1989.

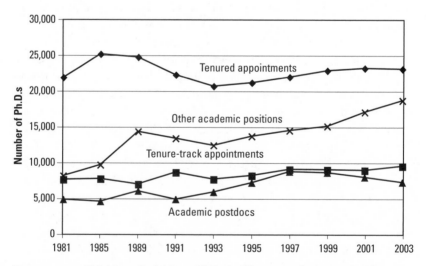

FIGURE 12.7 U.S. biomedical science Ph.D.s holding tenured or tenure-track appointments in academe, 1981–2003.

Source: NSF, Survey of Doctorate Recipients, 2003.

Note: Scale change reflects change in survey frequency after 1989.

done by people in their thirties. In the same time period, those fifty-six years and older competing for these NIH grants has increased from 13.3 percent to 21.3 percent (Moore 2002). This trend reflects the age structure of academic faculty, which is also increasing. Although retirement at age sixty-five is no longer mandatory, and retirements appear to be postponed (Ashenfelter and Card 2001), most choose to retire within a few years after this. A wave of retirements may be expected in the coming decade, as the baby boomers retire from tenured positions in academia. Will the pool of postdoctorals be of adequate quality to fill these vacated faculty positions, or will many of the best and brightest have chosen career paths instead in other sectors such as industry? A related and important question is whether universities will choose to fill these positions with tenure-track faculty when the faculty do retire.

Industry

Industry, including pharmaceutical and biotech companies, is a major source of employment for biomedical scientists. Jobs held by biomedical Ph.D.s in industry have increased steadily, doubling from 15 percent in 1981 to 32 percent in 2003 (Garrison et al. 2003) (Figure 12.6). Salaries tend to be higher than in academe, and industrial settings tend to have the latest in scientific equipment and seemingly unlimited supplies, all of which is appealing to the bench scientist. Job stability, however, is not as great in industry as it is in academe.

Other Job Sectors

Other job sectors have been fairly constant over time. For example, 10 percent of the jobs held by biomedical Ph.D.s are in government settings (e.g., research in government laboratories like NIH), and this has changed little since 1977 (Garrison et al. 2003) (Figure 12.6). Other jobs, such as those in patent law or science journalism, have also been constant over time but might be expected to increase as a byproduct of the biotech industry and its impact on our society. It can be questioned whether other jobs such as the latter need biomedical research training through the Ph.D., or whether a master's degree in the biomedical sciences would be adequate. It should also be noted that there has been less than 2 percent unemployment among individuals trained in biomedical sciences since 1977 (Garrison and Gerbi 1998; Garrison et al. 2003). This, of course, does not mean that the biomedical Ph.D. has obtained the type of employment that was desired or that they have fulfilled their career expectations.

CONCLUSION: CAN WE MEET THE BIOMEDICAL RESEARCH WORKFORCE NEEDS IN ACADEMIA?

As stated above, the biomedical research workforce in an academic laboratory relies heavily on predoctorals and postdoctorals. Increases in the NIH budget and the trend to larger laboratory groups for "big science" create an even greater demand for highly qualified laboratory personnel. As discussed above, the increase in biomedical Ph.D. production has not only leveled off but is now declining. Similarly, the number of U.S.-trained biomedical Ph.D.s who pursue postdoctoral training is declining, and the length of the postdoctoral period is also shortening. In light of these changes, how can we fill these workforce needs?

How should the scientific workforce be comprised? What is best for the research enterprise is not always best for the individual. The research enterprise benefits from the most highly trained, motivated, and intelligent laboratory personnel at the lowest cost to the system. In the recent past, such personnel have been graduate students and postdoctorals. A stable and experienced scientific team is an asset to the research project, but a lengthy predoctoral and postdoctoral period is not necessarily in the students' best interests.

With the shrinking number of graduate students and U.S.-trained predoctorals, how can we meet the expanding demand for biomedical research personnel in the academic laboratory? Much of the slack is currently taken up by an influx of postdoctorals who received their Ph.D.s in other countries (Garrison et al. 2005) (Figure 12.5). We are fortunate to be able to attract this group, but their training is not always comparable to that in a U.S. school, and they have often been brought up in a scientific culture different from ours. Difficulties with the English language can also create problems for them in oral and written communication. If they remain in the United States after their postdoctorals, they swell the numbers of people looking for permanent employment in the biomedical sciences. If, instead, they return to their home country, they may become industrial competitors.

We need to look beyond the current scheme of graduate students and postdoctorals to staff the academic research laboratory. Academia can learn from industry in this regard, as the latter manages without graduate students and only rarely has postdoctorals. The industrial workforce for biomedical research employs research assistants/technicians and Ph.D. staff scientists. It would be wise for academia to recognize the research scientist position for senior postdoctorals as an honorable career outcome with salaries commensurate with their experience (Marincola and Solomon 1998a and 1998b; Gerbi et al. 2001). Although research scientists will cost more than postdoctorals, they will add stability and productivity to the research enterprise.

Progress in biomedical research requires well-qualified laboratory personnel. Attracting good people requires promising career opportunities at the end of the training process. An integrated approach should be taken in designing training programs and in developing career opportunities that meet both the interests and concerns of students and the personnel needs of academic laboratories.

Concluding Remarks

Looking to the Future

PAULA E. STEPHAN AND RONALD G. EHRENBERG

In the three years since the conference that led to this volume, several changes have occurred relating to the practice of science at universities and the environment in which researchers and administrators are operating. For example, the dramatic decrease in the number of noncitizens accepted to graduate schools that occurred soon after 9/11 has diminished (National Academies 2005b). The pattern that had begun to emerge among postdoctoral fellows in the late 1990s of holding the postdoctoral positions for a shorter period of time has shown signs of reversal (Stephan and Ma 2005). Likewise, the trend among newly minted Ph.D.s to be less likely to take a postdoctoral position shows signs of reversal. Both of the latter reversals are likely related to poor job prospects in industry and academe that were occasioned by the recession at the turn of the century.

But much has not changed. Universities persist in embracing technology transfer as a means of facilitating economic growth and overcoming their budget problems. The inclination to press for earmarks continues. There is every indication that start-up packages remain large. The trend of hiring part-time and full-time non-tenure-track faculty rather than tenure-track faculty persists. And the resource gap between private and public universities continues to grow.

To the extent they continue, many of these trends have the potential of changing the quality of university science as well as the very fabric of the university experience. Continued earmarks are, as we have seen, associated with lower-quality research. Large start-up packages are one of several factors that lead universities to substitute non-tenure-track faculty for tenure-track faculty. While the stars that are hired with such packages are

259

highly productive, the system does much to create a chasm between the stars and the non-stars. Moreover, the failure to hire tenure-track faculty at the rate they were hired in the past means that there are fewer independent faculty researchers being hired and thus fewer investigator-initiated research projects being undertaken, especially by young investigators. Indeed, the latter was of sufficient concern that the National Academy of Sciences (2005a) issued a report that suggested strategies to increase the number of investigator-initiated awards in biomedical research going to individuals under thirty-five.

The debate continues as to whether the scientific commons are at risk of being cannibalized by the scramble of universities to translate research into new technologies, reaping the financial benefits in the process. Some argue that the changing university culture leads scientists and engineers increasingly to choose to allocate their time to research of a more applied as opposed to basic nature. Others express concern that the lure of economic rewards encourages scientists and engineers (and the universities where they work) to seek intellectual property (IP) protection for their results, eschewing (or postponing) publication and thus public disclosure. A related concern is that the granting of intellectual property hinders the ability of researchers to build on a given piece of knowledge.

The findings reported by the Thursbys in this volume, as well as their work elsewhere, suggest that faculty have not substituted applied research for basic research. Other scholars—for example, Azoulay, Ding, and Stuart (2004), Markiewicz and DiMinin (2004), and Stephan et al. (2007)—find patents and publications to be complementary activities. This, of course, does not address the concern with regard to postponement; nor does it address whether the granting of intellectual property hinders the ability of researchers to build on a given piece of knowledge. These are complex issues that require longitudinal data and careful analytical techniques. Moreover, in evaluating the findings, it is important to realize that there is considerable evidence that the commons were never governed under the Mertonian norms that have been so idealized by the critics of the consequences of technology transfer.

Even if the move to commercialization were benign in terms of its productivity effects, it does not follow that it has not come without costs. The goal of developing a strong technology transfer program is much like the goal of building a strong football team. The program is expensive, and only few, as Ehrenberg and coauthors remind us, reap sufficient rewards to even cover the cost of the technology transfer office. In a university environment where resources are scarce, one must question whether this is a prudent use of resources across all campuses. Moreover, there is indication that universities may have been overly successful in selling their contribution to innovation and economic growth to state legislatures. One consequence of this is that legislatures—witness the situation in California—have become convinced that the solution to local economic development is to create universities throughout the state. In eras of tight resources, this can lead to a serious deterioration in quality.

Looking ahead, the key question facing academic science and engineering may not be the degree to which commercialization negatively impacts scholarship in science and engineering. Instead, the question may be the extent to which changing personnel practices affect the quality of research performed in the university and the educational experience of students. To the extent that the university increasingly replaces tenure-track faculty with full-time and part-time non-tenure-track faculty, and to the extent postdoctoral fellows end up in dead-end careers, the enterprise is in danger of killing the goose that laid the golden egg.

Attracting talented young people into careers in science and engineering, and then providing opportunities for them to become researchers on their own merits, has always been the implicit contract of U.S. science. This contract, which began to experience considerable stress in the late 1980s and 1990s, is under even more stress today. It is no surprise that the number of citizen graduate students has declined in recent years. Nor is it surprising that the number of postdoctoral associations that focus on improving the working conditions and career opportunities for their members has increased in recent years. A strong academic science and engineering enterprise is important for the nation's well-being, and institutional and governmental leaders, as well as policy makers, would do well to contemplate the issues that we have raised here.

REFERENCES
CONTRIBUTORS
INDEX

References

Adams, J. D. 1990. Fundamental Stocks of Knowledge and Productivity Growth. *Journal of Political Economy* 98(4):673–702.

Adams, J. D., R. Clemmons, G. Black, and P. Stephan. 2002. Patterns of Research Collaboration in U.S. Universities. Mimeo, University of Florida, Gainesville.

Agrawal, A., and R. Henderson. 2002. Putting Patents in Context: Exploring Knowledge Transfer from MIT. *Management Science* 48(1):44–60.

Ainsworth, S. 1993. Regulating Lobbyists and Interest Group Influence. *Journal of Politics* 55:41–56.

Albert, M. B., D. Avery, F. Narin, and P. McAllister. 1991. Direct Validation of Citation Counts as Indicators of Industrially Important Patents. *Research Policy* 20(3):251–59.

America's Best Colleges. 2003. Washington, DC: U.S. News and World Report.

Anagnoson, J. T. 1980. Politics in the Distribution of Federal Grants: The Case of the Economic Development Administration. In *Political Benefits: Empirical Studies of American Public Programs,* ed. Barry S. Rundquist, 61–91. Lexington, MA: Lexington Books.

Andrikopoulos, A., J. Brox, and E. Carvalho. 1990. Shift-share Analysis and the Potential for Predicting Regional Growth Patterns: Some Evidence for the Region of Quebec, Canada. *Growth and Change* 21(1):1–10.

Arnold, R. D. 1981. Legislators, Bureaucrats, and Locational Decisions. *Public Choice* 37:107–32.

Arnone, M. 2004. Security at Home Creates Insecurity Abroad. *Chronicle of Higher Education* 50, no. 27 (March 12): A21–22.

Ashenfelter, O., and D. Card. 2001. Did the Elimination of Mandatory Retirement Affect Faculty Retirement Flows? National Bureau of Economic Research Working Paper 8378.

Association of University Technology Managers, Inc. 2000. *AUTM Licensing Survey: FY 1999.* Northbrook, IL: Association of University Technology Managers.

Association of University Technology Managers, Inc. 2001. *AUTM Licensing Survey: FY 2000.* Northbrook, IL: Association of University Technology Managers.

Attiyeh, G., and R. Attiyeh. 1997. Testing for Bias in Graduate School Admissions. *Journal of Human Resources* 32(3):2997.

Audretsch, D., and P. Stephan. 1996. Company-scientist Locational Links: The Case of Biotechnology. *American Economic Review* 86(3):641–52.

Audretsch, D., and M. Feldman. 1996a. Innovation Clusters and the Industry Life Cycle. *Review of Industrial Organization* 11(2):253.

Audretsch, D., and M. Feldman. 1996b. R&D Spillovers and the Geography of Innovation and Production. *American Economic Review* 63(3):30.

Austen-Smith, D. 1993. Information and Influence: Lobbying for Agendas and Votes. *American Journal of Political Science* 37(3):799–834.

———. 1995. Campaign Contributions and Access. *American Political Science Review* 89(3):566–81.

Azoulay, P., W. Ding, and T. Stuart. 2004. The Impact of Academic Patenting on (Public) Research Output. Unpublished paper.

Barber, B. 1952. *Science Social Order.* Glencoe, IL: Free Press.

Barff, R. A., and P. L. Knight III. 1988. Dynamic Shift-share Analysis. *Growth and Change* 19(2):1–10.

Barron's Profiles of American Colleges. Various years. Hauppauge, NY: Barron's Educational Series, Inc.

Barton, J. 1998. The Impact of Contemporary Patent Law on Plant Biotechnology Research. In *Global Genetic Resources: Access and Property Rights,* ed. S. A. Eberhart et al., 85–97. Proceedings of the Intellectual Property Rights III Conference. Madison, WI: Crop Science Society of America.

Best and Brightest Avoiding Science. 2000. *Science* 288(5463):43.

Betts, J. R. 1998. Educational Crowding Out: Do Immigrants Affect the Educational Attainment of American Minorities. In *Help or Hindrance: The Economic Implications of Immigration for African Americans,* ed. D. S. Hamermesh and F. D. Bean, 253–81. New York: Russell Sage Foundation.

Bhagwati, J., and M. Rao. 1996. The False Alarm of "Too Many Scientists." *The American Enterprise* 7:71–72.

Black, G. 2004. *The Geography of Small Firm Innovation.* New York: Kluwer.

Blumberg, P. 1996. From "Publish or Perish" to "Profit or Perish": Revenues from University Technology Transfer and the § 501 (c) (3) Tax Exemption. *University of Pennsylvania Law Review* 145:89.

Blumenstyk, G. 2001. A Vilified Corporate Partnership Produces Little Change (Except Better Facilities). *Chronicle of Higher Education* 47, no. 41 (June 22): A24–27.

———. 2002a. Income from University Licenses on Patents Exceeded $1-Billion. *Chronicle of Higher Education* (March 22): A31.

———. 2002b. U. of Rochester Risks Millions in Patent Fight with Pharmaceutical Giants. *Chronicle of Higher Education* (September 20): A27.

———. 2002c. Value of University Licenses on Patents Exceeded $1-Billion in 2000, Survey Finds. *Chronicle of Higher Education* (March 5): A21.

———. 2003a. Federal Court Dismisses U. of Rochester's Suit that Sought Billions for Patent Infringement. *Chronicle of Higher Education* (March 21): A31.

———. 2003b. University Earnings from Inventions Hit Nearly $1-Billion in 2002, Survey Finds. *Chronicle of Higher Education* (December 9): A28.

———. 2003c. A New Johnny Appleseed. *Chronicle of Higher Education* 49 (July 18): A25.

Blumenthal, D. 2002. Conflict of Interest in Biomedical Research. *Health Matrix* 12:377–92.

———. 2003. Academic-Industrial Relationships in the Life Sciences. *New England Journal of Medicine* 349(25):2452–59.

Blumenthal, D., E. G. Campbell, M. S. Anderson, N. Causino, and K. S. Louis. 1997. Withholding Research Results in Academic Life Science: Evidence from a National Survey of Faculty. *Journal of the American Medical Association* 277(15):1224–28.

Blumenthal, D., N. Causino, E. Campbell, and K. S. Louis. 1996. Relationships Between Academic Institutions and Industry in the Life Sciences. *New England Journal of Medicine* 334(6):368–74.

Blumenthal, D., M. Gluck, K. S. Louis, M. A. Stoto, and D. Wise. 1986. University-Industry Research Relationships in Biotechnology: Implications for the University. *Science* 232(4756): 1361–66.

Bok, D. 2003. Universities in the Marketplace: The Commercialization of Higher Education. Princeton, NJ: Princeton University Press.

Borjas, G. J. 1994. The Economics of Immigration. *Journal of Economic Literature* 32(4):1667–1717.

———. 1995. The Economic Benefits from Immigration. *Journal of Economic Perspectives* 9(2):3–22.

———. 2000. Foreign-born Teaching Assistants and the Academic Performance of Undergraduates. *American Economic Review* 90(2):355–59.

———. 2002. *An Evaluation of the Foreign Student Program.* Washington, DC: Center for Immigration Studies.

———. 2003. The Labor Demand Curve *Is* Downward Sloping: Reexamining the Impact of Immigration on the Labor Market. *Quarterly Journal of Economics* 118(4):1335–74.

Bouvier, L., and Martin, J. 1995. *Foreign-born Scientists, Engineers and Mathematicians in the United States.* Washington, DC: Center for Immigration Studies.

Bowen, W. G., and N. L. Rudenstine. 1992. *In Pursuit of the PhD.* Princeton, NJ: Princeton University Press.

Bowen, W. G., S. E. Turner, and M. L. Witte. 1992. The B.A.-Ph.D. Nexus. *Journal of Higher Education* 63(1):65–86.

Bowie, N. 1994. *University-Business Partnerships: An Assessment.* Lanham, MD: Rowman and Littlefield.

Brainard, J. 2002. Some Colleges Reap Little Return Lobbying for Pork Projects, Study Finds. *Chronicle of Higher Education* (October 18): A27.

———. 2005. The Ghosts of Stanford: Have Radical Constraints on Reimbursing Overhead for Research Grants Gone Too Far? *Chronicle of Higher Education* 51 (August 5): A16–18.

Brainard, J., and A. M. Borrego. 2003. Academic Pork Barrel Tops $2-Billion for First Time. *Chronicle of Higher Education* (September 26): A18.

Bragdon, S. H., and D. R. Downes. 1998. Recent Policy Trends and Developments Related to the Conservation, Use and Development of Genetic Resources. *Issues in Genetic Resources* 7 (June): 1–42.

Bremer, H. W. 2001. The First Two Decades of the Bayh-Dole Act as Public Policy. Presentation to National Association of State Universities and Land Grant Colleges (November 11), Washington, DC.

Brief for Respondents. 2006. *Laboratory Corporation v. Metabolite Laboratories*. U.S. S. Ct. Briefs LEXIS 220.

Broad, W. J. 2004. U.S. is Losing its Dominance in the Sciences. *New York Times* (May 3): A1.

Burchfiel, K. J. 1995. *Biotechnology and the Federal Circuit*. Washington, DC: Bureau of National Affairs.

Busch, L., et al. 2004. External Review of the Collaborative Research Agreement between Novartis Agricultural Discovery Institute, Inc. and the Regents of the University of California. http://www.berkeley.edu/news/media/releases/2004/07/external_novartis_review.pdf.

Butler, L. 2002. A List of Published Papers is No Measure of Value. *Nature* 419 (October 13): 877.

Caldart, C. C. 1983. Industry Investment in University Research. *Science, Technology, and Human Values* 8(2):24–32.

Carnegie Commission on Higher Education. 1994. *A Classification of Institutions of Higher Education*. Princeton, NJ: Carnegie Foundation for the Advancement of Teaching.

Carvajal, D. 2004. Reversing the Expat Brain Drain. *International Herald Tribune* (March 10): 15.

Catholic U. Pulls Out of the Association of American Universities. 2002. *Chronicle of Higher Education* 49 (November 15): A23.

Cho, M. K., and L. A. Bero. 1996. The Quality of Drug Studies Published in Symposium Proceedings. *Annals of Internal Medicine* 124(55):485–89.

Chubin, D. E., and E. J. Hackett. 1990. *Peerless Science: Peer Review and U.S. Science Policy*. Albany: State University of New York Press.

Clayton, M. 2001. Corporate Cash and Campus Labs. *Christian Science Monitor* (June 19): 11.

Coffman, W. R., J. E. Alexander, D. J. BenDaniel, P. L. Carey, H. G. Craighead, C. R. Fay, P. A. Gould, J. S. Gross, J. E. Hunter, W. H. Lesser, S. Loker, J. N. MacLeod,

J. J. Mingle, N. R. Scott, and A. F. Krattiger. 2003. The Future of Technology Transfer at a Major Land Grant University: Report of the Cornell University Land Grant Panel on Technology Transfer. *IP Strategy Today* 6:1–41.

Cohen, W., R. Florida, L. Randazzese, and J. Walsh. 1998. Industry and the Academy: Uneasy Partners in the Cause of Technological Advance. In *Challenges to Research Universities*, ed. R. Noll, 171–99. Washington, DC: Brookings Institution Press.

Cohen, W., R. Florida, and R. Goe. 1994. *University-Industry Research Centers in the United States.* Report to the Ford Foundation. Pittsburgh, PA: Carnegie Mellon University.

Colyvas, J., M. Crow, A. Gelijns, R. Mazzoleni, R. Nelson, N. Rosenberg, and B. Sampat. 2002. How Do University Inventions Get into Practice? *Management Science* 48(1):61–72.

Cook, P. J., and R. H. Frank. 1993. The Growing Concentration of Top Students at Elite Schools. In *Studies of Supply and Demand in Higher Education,* ed. C. T. Clotfelter and M. Rothschild, 121–40. Chicago: University of Chicago Press.

Cotropia, C. 2006. Observations on Recent Patent Decisions: The Year in Review. *Journal of the Patent and Trademark Office Society* 88(1):46.

Cowley, J. M., and R. Makowski. 2003. Back to the Future: Rethinking the Product of Nature Doctrine as a Barrier to Biotechnology Patents (Part 1). *Journal of the Patent and Trademark Office Service* 85:301–34.

Cripps, Y. 2004. The Art and Science of Genetic Modification: Re-Engineering Patent Law and Constitutional Orthodoxies. *Indiana Journal of Global Legal Studies* 11(1):1–30.

Dechenaux, E., B. Goldfarb, S. Shane, and M. Thursby. 2005. Appropriability and the Timing of Innovation: Evidence from MIT Inventions. National Bureau of Economic Research Working Paper 9725.

de Figueiredo, John M., and Brian Silverman. 2006. Academic Earmarks and the Returns to Lobbying. *Journal of Law and Economics* 49(2):597–626.

de Figueiredo, J. M., and E. H. Tiller. 2001. The Structure and Conduct of Corporate Lobbying: An Empirical Analysis of Corporate Lobbying at the Federal Communications Commission. *Journal of Economics and Management Strategy* 10(1):91–122.

Demaine, L. J., and A. X. Fellmeth. 2003. Patent Law: Natural Substances and Patentable Inventions. *Science* 300(5624):1375–76.

Dembner, A. 2003. Rallies Aim to Save Canadian Drug Sales. *Boston Globe* (February 20) A1.

Diamond v. Chakrabarty. 1980. 447 U.S. 303.

Diamond v. Diehr. 1981. 450 U.S. 175.

Dickson. D. 1988. *The New Politics of Science.* New York: Pantheon Books.

Dueker, K. S. 1997. Biobusiness on Campus: Commercialization of University-Developed Biomedical Technologies. *Food and Drug Law Journal* 52(4):453–509.

Ehrenberg, R. G. 1992. The Flow of New Doctorates. *Journal of Economic Literature* 30(2):830–75.

———. 2000. *Tuition Rising: Why College Costs So Much*. Cambridge, MA: Harvard University Press.

———. 2003. Studying Ourselves: The Academic Labor Market. *Journal of Labor Economics* 21:267–87.

Ehrenberg, R. G., and J. Epifantseva. 2001. Has the Growth of Science Crowded Out Other Things at Universities? *Change* 33(4): 46–52.

Ehrenberg, R. G., D. B. Klaff, A. T. Kezsbom, and M. P. Nagowski. 2004. Collective Bargaining in American Higher Education. In *Governing Academia*, ed. R. G. Ehrenberg, 209–34. Ithaca, NY: Cornell University Press.

Ehrenberg, R. G., M. J. Rizzo, and G. H. Jakubson. 2003. Who Bears the Growing Cost of Science at Universities? National Bureau of Economic Research Working Paper 9627.

Ehrenberg, R. G., and C. L. Smith. 2003. The Sources and Uses of Annual Giving at Selective Private Research Universities and Liberal Arts Colleges. *Economics of Education Review* 22(3):223–35.

Ehrenberg, R. G., and L. Zhang. 2005. The Changing Nature of Faculty Employment. In *Recruitment, Retention, and Retirement in Higher Education: Building and Managing the Faculty of the Future*, ed. R. Clark and J. Ma, 32–52. Northampton, MA: Edward Elgar Publishers.

Eisenberg, R. S. 1987. Proprietary Rights and the Norms of Science in Biotechnology Research. *Yale Law Journal* 97(2):177–231.

———. 1988. Academic Freedom and Academic Values in Sponsored Research. *Texas Law Review* 66:1363–1404.

———. 1989. Patents and the Progress of Science: Exclusive Rights and Experimental Use. *University of Chicago Law Review* 56:1017–86.

———. 1996. Public Research and Private Development: Patents and Technology Transfer in Government-Sponsored Research. *Virginia Law Review* 82:1663–1727.

Elliott, V. 2001. Who Calls the Tune? The Money Game: Novartis Strategic Alliance with the University of California at Berkeley. *UNESCO Courier* (November 1): 21.

Enríquez, J. 1998. Genomics and the World Economy. *Science* 281(5379):925–26.

Etzkowitz, H., A. Webster, and P. Healey, eds. 1998. *Capitalizing Knowledge: New Intersections of Industry and Academia*. Albany: State University of New York Press.

Fang, D., and R. E. Meyer. 2003. Ph.D. Faculty in Clinical Departments of U.S. Medical Schools, 1981–1999: Their Widening Presence and Roles in Research. *Academic Medicine* 78:167–76.

Ferejohn, J. A. 1974. *Pork Barrel Politics: Rivers and Harbors Legislation, 1947–1968*. Stanford, CA: Stanford University Press.

Field, K. 2004. NSF Eliminates Cost Sharing. *Chronicle of Higher Education* 51 (October 29): A27.

Finn, M. G. 1997. Stay Rates of Foreign Doctorate Recipients from U.S. Universities, 1995. Oak Ridge, TN: Oak Ridge Institute for Science and Education.

Finn, M. G. 2000. Stay Rates of Foreign Doctorate Recipients from U.S. Universities, 1997. Oak Ridge, TN: Oak Ridge Institute for Science and Education.

Finn, M. G. 2003. Stay Rates of Foreign Doctorate Recipients from U.S. Universities, 2001. Oak Ridge, TN: Oak Ridge Institute for Science and Education.

Finn, R. 1995. Scientists' Heated Debate on Immigration Mirrors Issues Argued Throughout U.S. *The Scientist* 9(23):1, 8–9.

Fleisher, B., M. Hashimoto, and B. Weinberg. 2002. Foreign GTAs Can Be Effective Teachers of Economics. *Journal of Economic Education* 33(4):299–325.

Frank, L. 2001. A Biotech Gambit in the Desert. *Science* 292(5521):1478.

Freeman, R. B., E. Weinstein, E. Marincola, J. Rosenbaum, and F. Solomon. 2001. Competition and Careers in Bioscience. *Science* 294(5550):2293–94.

Frey, K. J. 1996. National Plant Breeding Study–I: Human and Financial Resources Devoted to Plant Breeding Research and Development in the United States in 1994. Special report 98. Ames: Iowa Agricultural and Home Economics Experiment Station, Iowa State University.

———. 1997. National Plant Breeding Study–II: National Plan for Promoting Breeding Programs for Minor Crops in the United States. Special report 100. Ames: Iowa Agricultural and Home Economics Experiment Station, Iowa State University.

———. 1998. National Plant Breeding Study–III: National Plan for Genepool Enrichment of U.S. Crops. Special report 101. Ames: Iowa Agricultural and Home Economics Experiment Station, Iowa State University.

Garde, T. 2005. Supporting Innovation in Targeted Treatments: Licenses of Right to NIH-Funded Research Tools. *Michigan Telecommunications and Technology Law Review* 11: 249–84.

Garrison, H. H., and S. A. Gerbi. 1998. Education and Employment Patterns of U.S. Ph.D.s in the Biomedical Sciences. *FASEB Journal* 12:139–48.

Garrison, H. H., S. A. Gerbi, and P. W. Kincade. 2003. In an Era of Scientific Opportunity, Are There Opportunities for Biomedical Scientists? *FASEB Journal* 17:2169–73.

Garrison, H. H., A. L. Stith, and S. A. Gerbi. 2005. Foreign Postdocs: The Changing Face of Biomedical Science in the U.S. *FASEB Journal* 19:1938–42.

Gerbi, S. A., H. H. Garrison, and J. P. Perkins. 2001. Workforce Alternatives to Graduate Students? *Science* 292(5521): 1489–90.

Geiger, R., and I. Feller. 1995. The Dispersion of Academic Research in the 1980s. *Journal of Higher Education* 66(3):336–60.

Gibbons, M., C. Limoges, H. Nowotny, S. Schwartzman, P. Scott, and M. Trow. 1994. The New Production of Knowledge: The Dynamics of Science and Research in Contemporary Societies. London: Sage.

Glanz, J. 2001. The World of Science Becomes a Global Village: Archive Opens a New Realm of Research. *New York Times* (May 1): D1.

Goldin, C., and L. F. Katz. 1998. The Origins of State-level Differences in the Public Provision of Higher Education: 1890–1940. *The American Economic Review* 88:303–8.

———. 1999. The Shaping of Higher Education: The Formative Years in the United States, 1890 to 1940. *Journal of Economic Perspectives* 13:37–62.

Goldman, C. A., and W. F. Massy. 2001. *The PhD Factory: Training and Employment of Science and Engineering Doctorates in the United States*. Bolton, MA: Anker.

Gordon, J., W. Hackett, and D. Mulkey. 1980. Using the Shift-share Technique in Economies with Widely Varying Sectoral Growth Rates: Observations and a Suggested Model Modification. *The Review of Regional Studies* 10(1):57–67.

Gray, W. H. 1994. Pork or Providence? A Defense of Earmarked Funds for Colleges. *Washington Post* (April 27): A23.

Greenspan Mulls Patent Problems. 2003. BBC News. http://news.bbc.co.uk/go/pr/fr/ -/1/business/2918691.stm. Published 2003/04/04 17:21:07.

Griffin, G. 2006. CU Profs Taking Patent to Justices. *Denver Post* (March 20): C1.

Grobar, L. 1996. Comparing the New England and Southern Regional Recessions. *Contemporary Economic Policy* 14:71–84.

Groen, J. A., and M. J. Rizzo. 2003. The Changing Composition of American-Citizen PhDs. Paper presented at Science and the University conference, Cornell Higher Education Research Institute, Ithaca, NY, May 20–21. http://www.ilr.cornell.edu/ cheri/.

Hagstrom, W. 1965. *The Scientific Community*. New York: Basic Books.

Hall, B., A. Jaffe, and M. Trajtenberg. 2001. The NBER Patent Citation Data File: Lessons, Insights and Methodological Tools. National Bureau of Economic Research Working Paper 8498.

Hall, R. L., and B. Grofman. 1990. The Committee Assignment Process and the Conditional Nature of Committee Bias. *American Political Science Review* 84(4):1149–66.

Hardy, K. R. 1974. Social Origins of American Scientists and Scholars. *Science* 185(4150):497–506.

Harmon, L. R. 1978. *A Century of Doctorates: Data Analysis of Growth and Change*. Washington, DC: National Academy of Sciences.

———. 2001. *Patents and the Federal Circuit*, 5th ed. Washington, DC: Bureau of National Affairs.

Hawk, J. A., and M. E. Smith. 1993. The Role of Corn Breeding in Future Northeastern Crop Production. In *Agricultural Research in the Northeastern United States: Critical Review and Future Perspectives*, ed. T. J. Sims, 81–93. Madison, WI: American Society of Agronomy.

Haynes, K. E., and M. Dinc. 1997. Productivity Change in Manufacturing Regions: A Multifactor/Shift-share Approach. *Growth and Change* 28(2):201–21.

Heller, M. A., and R. S. Eisenberg. 1998. Can Patents Deter Innovation? The Anticommons in Biomedical Research. *Science* 280(5364):698–701.

Herdt, R. W. 1999. Enclosing the Global Plant Genetic Commons. Prepared for delivery at the China Center for Economic Research (May 24), based on a paper delivered at the Institute for International Studies, Stanford University, January 14. http://www.rockfound.org.

Hicks, D., A. Breitzman, K. Hamilton, and F. Narin. 2000. Research Excellence and Patented Innovation. *Science and Public Policy* 27(5):310–20.

Hicks, D., T. Breitzman, D. Olivastro, and K. Hamilton. 2001. The Changing Composition of Innovative Activity in the U.S.—A Portrait Based on Patent Analysis. *Research Policy* 30(4):681–703.

High Court Hears Patent Arguments. 2006. *Richmond Times Dispatch* (March 22): C2.

Hill, S. T. 1992. Undergraduate Origins of Recent Science and Engineering Doctorate Recipients. Washington, DC: National Science Foundation.

Hoffer, T. B., B. L. Dugoni, A. R. Sanderson, S. Sederstrom, R. Ghadialy, and P. Rocque. 2001. *Doctorate Recipients from United States Universities: Summary Report 2000.* Chicago: National Opinion Research Center.

Hofstadter, R., and W. P. Metzger. 1955. *The Development of Academic Freedom in the United States.* New York: Columbia University Press.

Hoppes, R. 1997. Shift-share Analysis for Regional Health Care Policy. *Journal of Regional Analysis and Policy* 27(1):35–45.

Hoxby, C. M. 1997. How the Changing Market Structure of U.S. Higher Education Explains College Tuition. National Bureau of Economic Research Working Paper W6323.

———. 1998. Do Immigrants Crowd Disadvantaged American Natives Out of Higher Education? In *Help or Hindrance? The Economic Implications of Immigration for African Americans,* ed. D. S. Hamermesh and F. D. Bean, 282–321. New York: Russell Sage Foundation.

Huvane, K. 2002. Researchers Required to Show Money Trail. *World Watch* 15(1):7.

In re Fisher. 2005. 421 F.3d 1365 (Fed. Cir.).

Institute of International Education. 2004. Open Doors Report on International Educational Exchange: Survey of Foreign Student and Scholar Enrollment and Visa Trends for all 2004. New York: Institute of International Education.

Ishikawa, Y. 1992. The 1970s Migration Turnaround in Japan Revisited: A Shift-share Approach. *Papers in Regional Science* 71(2):153–73.

Jacobs, L., and C. B. Friedman. 1988. Student Achievement Under Foreign Teaching Associates Compared with Native Teaching Associates. *Journal of Higher Education* 69(5):551–63.

Jaffe, A. 1989. Real Effects of Academic Research. *American Economic Review* 79(5):957–70.

James, C. 2002. Preview: Global Status of Commercialized Transgenic Crops: 2002. *ISAAA Briefs* No. 27. Ithaca, NY: ISAAA.

Jensen, R., and M. Thursby. 2001. Proofs and Prototypes for Sale: The Licensing of University Inventions. *American Economic Review* 91(1):240–59.

Jensen, R., J. Thursby, and M. Thursby. 2003. The Disclosure and Licensing of University Inventions: The Best We Can Do with the S**t We Get to Work With. *International Journal of Industrial Organization* 21:1271–1300.

Jensen, R., and M. Thursby. 2004. Patent Licensing and the Research University. National Bureau of Economic Research Working Paper 11128.

Johnson, G. E. 1998. Estimation of the Impact of Immigration on the Distribution of Income among Minorities and Others. In *Help or Hindrance? The Economic Implications of Immigration for African Americans,* ed. D. S. Hamermesh and F. D. Bean, 17–50. New York: Russell Sage Foundation.

Johnson, J.M. 1993. *Human Resources for Science and Technology: The Asian Region.* NSF 93-303. Washington, DC: National Science Foundation.

Johnson, J. M., and H. R. Coward. 2000. Summary of Proceedings. In *Graduate Education Reform in Europe, Asia, and the Americas and International Mobility of Scientists and Engineers: Proceedings of an NSF Workshop,* ed. J. M. Johnson, 2–3. NSF 00–318. Arlington, VA: National Science Foundation, Division of Science Resources Studies.

Johnston, J. 2007. Health Related Academic Technology Transfer: Rethinking Patenting and Licensing Practices. *International Journal of Biotechnology* 9(2):156–71.

Joughin, Louis. 1967. *Academic Freedom and Tenure: A Handbook of the American Association of University Professors.* Madison: University of Wisconsin Press.

Kapczynski, A., S. Chaifetz, Z. Katz, and Y, Benkler. 2005. Addressing Global Health Inequities: An Open Licensing Approach for University Innovations. *Berkeley Technology Law Journal* 20:1031.

Keil, S. 1992. On the Value of Hometheticity in the Shift-share Framework. *Growth and Change* 23(4):469–93.

Kennedy, D. 1997. *Academic Duty.* Cambridge, MA: Harvard University Press.

———. 2001. Good News, Bad News. *Science* 293(5531):761.

Kenney, M. 1986. *Biotechnology: The University-Industrial Complex.* New Haven, CT: Yale University Press.

Klein, T. M., E. D. Wolf, R. Wu, and J. C. Sanford. 1987. High-velocity Microprojectiles for Delivering Nucleic Acids into Living Cells. *Nature* 327(6117):70–73.

Kleinman, D. 2001. Systemic Influences: Some Effects of the World of Commerce on University Science. In *Degrees of Compromise: Industrial Interests and Academic Values,* ed. J. Croissant and S. Restivo, 225–40. Albany: State University of New York Press.

———. 2003. *Impure Cultures: University Biology and the World of Commerce.* Madison: University of Wisconsin Press.

———. 2005. *Science and Technology in Society: From Biotechnology to the Internet.* Malden, MA: Blackwell.

Kling, R., and G. G. McKim. 1999. Scholarly Communication and the Continuum of Electronic Publishing. *Journal of the American Society for Information Science* 50(10):890–906.

———. 2000. Not Just a Matter of Time: Field Differences and the Shaping of Electronic Media in Supporting Scientific Communication. *Journal of the American Society for Information Science* 51(14):1306–20.

Kloppenburg, J., Jr. 2004. *First the Seed: The Political Economy of Plant Biotechnology.* Madison: University of Wisconsin Press.

Knapp, R. H., and H. B. Goodrich. 1952. *Origins of American Scientists.* Chicago: University of Chicago Press.

Kocher, M. G., and M. Sutter. 2001. The Institutional Concentration of Authors in Top

Journals of Economics During the Last Two Decades. *The Economic Journal* 111:F405–21.

Krimsky, S. 1999. The Profit of Scientific Discovery and Its Normative Implications. *Chicago-Kent Law Review* 75(1):15–39.

Krimsky, S. 2003. *Science in the Private Interest.* Lanham, MD: Rowman & Littlefield.

Lach, J. S., and M. Schankerman. 2003. Incentives and Invention in Universities. Center for Economic Policy Research Working Paper 3916.

La Porta, R., F. Lopez-de-Silanes, A. Shleifer, and R. Visny. 1999. The Quality of Government. *Journal of Law, Economics, and Organization* 15(1):222–79.

Lawler, A. 2003. Last of the Big Time Spenders? *Science* 299(5605):330–33.

Leroux, C. 2001. Biotech's Traffic Cop: Chicago Attorney Lori Andrews Stands Where Science and the Law Intersect. *Chicago Tribune* (October 7): C12.

Leslie, S. 1993. *The Cold War and American Science: The Military-Industrial Academic Complex at MIT and Stanford.* New York: Columbia University Press.

Levin, S., and P. E. Stephan. 1991. Research Productivity Over the Life Cycle: Evidence for American Scientists. *American Economic Review* 81(1):114–32.

———. 1999. Are the Foreign Born a Source of Strength for U.S. Science? *Science* 285(5431):1213–14.

Levin, S., G. Black, A. Winkler, and P. Stephan. 2004a. Differential Employment Patterns for Citizens and Non-Citizens in Science and Engineering in the United States: Minting and Competitive Effects. *Growth and Change* 35(4):456–75.

———. 2004b. The Changing Career Outcomes of Citizen and Non-Citizen Scientists and Engineers in Higher Education. Mimeo, Georgia State University, Atlanta.

Lieberwitz, R. L. 2002. The Corporatization of the University: Distance Learning at the Cost of Academic Freedom? *Boston University Public Interest Law Journal* 12(1):73.

———. 2004. The Marketing of Higher Education: The Price of the University's Soul. Review of *Universities in the Marketplace: The Commercialization of Higher Education* by D. Bok. *Cornell Law Review* 89:763–800.

———. 2005. Confronting the Privatization and Commercialization of Academic Research: An Analysis of Social Implications at the Local, National, and Global Levels. *Indiana Journal of Global Legal Studies* 12:109–52.

Liu, M., and W. T. Mallon. 2004. Tenure in Transition: Trends in Basic Science Faculty. *Academic Medicine* 79(3):205–13.

Lomperis, A. M. T. 1992. The Demographic Transformation of American Doctoral Education. *Research in Labor Economics* 13:131–213.

Madey v. Duke University. 2002. 307 F.3d 1351 (Fed. Cir. 2002).

Mansfield, E. 1995. Academic Research Underlying Industrial Innovations: Sources, Characteristics, and Financing. *The Review of Economics and Statistics* 77(1):55–65.

Marincola, E., and F. Solomon. 1998a. The Career Structure in Biomedical Research: Implications for Training and Trainees. *Molecular Biology of the Cell* 9:3003–6.

———. 1998b. Training for Today's Marketplace. *Science* 281(5377):645.

Markiewicz, K., and A. DiMinin. 2004. Commercializing the Laboratory: The

Relationship Between Faculty Patenting and Publishing. Unpublished paper. University of California, Berkeley.

Matloff, N. 1988. Debunking the Myth of a Desperate Software Labor Shortage. Testimony to the U.S. House Judiciary Committee on Immigration (updated April 8 1988).

Mathae, K., and C. Bitzer, eds. 2004. *Reinvigorating the Humanities: Enhancing Research and Education on Campus and Beyond.* Washington, DC: Association of American Universities.

Mayer, K. R. 1991. *The Political Economy of Defense Contracting.* New Haven, CT: Yale University Press.

McConnell, R. 1997. Public and Private Corn Breeding: Developing Partnerships for the Future. In Proceedings of the North Central Corn Breeding Research Committee (NCR-167). Ames: Iowa State Agricultural Experiment Station, Iowa State University.

McCouch, S. R., G. Kochert, Z. Yu, Z. Wang, G. S. Khush, W. R. Coffman, and S. D. Tanksley. 1988. Molecular Mapping of Rice Chromosomes. *Theoretical and Applied Genetics* 76(6):815–29.

Merck KGAA v. Integra Lifesciences I, Ltd. 2005. 545 U.S. 193.

Merton, Robert K. 1973. *The Sociology of Science.* Chicago: University of Chicago Press.

Metabolite Labs., Inc., v. Lab. Corp. of Am. Holdings. 2006. 370 F.3d 1354 (Fed. Cir. 2004), cert. granted in part, 126 S.Ct. 601 (2005), cert. dismissed as improvidently granted, 2006 U.S. LEXIS 4893.

Mikhail, P. 2000. Hopkins v. CellPro: An Illustration that Patenting and Exclusive Licensing of Fundamental Science Is Not Always in the Public Interest. *Harvard Journal of Law and Technology* 13(2):375.

MIT. 2000. MIT's Alliances with Industry. *MIT News:* http://web.mit.edu/newsoffice/nr/2000/alliance.html (last accessed April 16, 2007).

Moore, R. 2002. Personal Communication. Office of Extramural Research, National Institutes of Health, Bethesda, MD.

Morgan, R., N. Kannankutty, and D. Strickland. 1997. Future Directions for University-Based Engineering Research. *ASEE Prism* 6(7):33–36.

Mowery, D. C., and A. Ziedonis. 2002. Academic Patent Quality and Quantity Before and After the Bayh-Dole Act in the United States. *Research Policy* 31(3):399–418.

Mowery, D. C., R. Nelson, B. Sampat, and A. Ziedonis. 1999. The Effects of the Bayh-Dole Act on U.S. University Research and Technology Transfer: An Analysis of Data from Columbia University, the University of California, and Stanford University. In *Industrializing Knowledge: University-Industry Linkages in Japan and the United States,* ed. L. M. Branscomb, F. Kodama, and R. Florida, 269–306. Cambridge, MA: MIT Press.

Mowery, D. C., R. Nelson, B. Sampat, and A. Ziedonis. 2001. The Growth of Patenting and Licensing by U.S. Universities: An Assessment of the Effects of the Bayh-Dole Act of 1980. *Research Policy* 30(1):99–119.

———. 2004. *Ivory Tower and Industrial Innovation: University-Industry Technology Transfer Before and After the Bayh-Dole Act.* Stanford, CA: Stanford Business Books.

Mowery, D. C., and N. Rosenberg. 1993. The U.S. National Innovation System. In *National Innovation Systems*, ed. R. R. Nelson, 29–75. New York: Oxford University Press.

Murphy, K. M., A. Shleifer, and R. W. Vishny. 1993. Why is Rent-Seeking So Costly to Growth? *American Economic Review* 83(2):409–14.

Murray, F. 2002. Innovation as Coevolution of Science and Technology: Exploring Tissue Engineering. *Research Policy* 31(8–9):1389–403.

———. 2006. The Oncomouse that Roared: Resistance and Accommodation to Patenting in Academic Science. Unpublished paper.

Narin, F. 1976. Evaluative Bibliometrics: The Use of Publication and Citation Analysis in the Evaluation of Scientific Activity. Contract NSF C-627. National Science Foundation (March 31).

Narin, F., G. Pinski, and H. Gee. 1976. Structure of the Biomedical Literature. *Journal of the American Society for Information Science* 27(1):25–45.

National Academy of Sciences. 1958. *Doctorate Production in United States Universities, 1936–1956, with Baccalaureate Origins of Doctorates in Sciences, Arts and Humanities.* Publication 582. Washington, DC: National Academy of Sciences, National Research Council.

National Academy of Sciences. 1967. *Doctorate Recipients from United States Universities, 1958–1966.* Publication 1489. Washington, DC: National Academy of Sciences.

National Academy of Sciences. 2005a. *Bridges to Independence: Fostering the Independence of New Investigators in Biomedical Research.* Washington, DC: National Academies Press.

National Academy of Sciences. 2005b. *Policy Implications of International Graduate Students and Postdoctoral Scholars in the United States.* Washington, DC: National Academies Press.

National Center for Education Statistics. 2002. *Digest of Education Statistics 2001.* Washington, DC: U.S. Department of Health, Education, and Welfare, Office of Education.

National Center for Education Statistics. Various years. *Earned Degrees Conferred.* Washington, DC: U.S. Department of Health, Education, and Welfare, Office of Education.

National Center for Education Statistics. Various years. Higher Education General Information Survey (HEGIS).

National Center for Education Statistics. Various years. Integrated Postsecondary Education Data System (IPEDS). Data available on WebCASPAR. http://caspar.nsf.gov.

National Institutes of Health. 1999. Principles and Guidelines for Recipients of NIH Research Grants and Contracts on Obtaining and Disseminating Biomedical Research Resources: Final Notice. *64 Fed. Reg. 72,090* (December 23).

National Institutes of Health. 2003. NIH Calls on Scientists to Speed Public Release of Research Publications. *NIH News.* www.hih.gov/news/pr/feb2005/od-03.htm (last accessed June 7, 2007).

National Institutes of Health. 2004. Best Practices for the Licensing of Genomic Inventions. *69 Fed. Reg. 67,747* (November 19).

National Research Council. 1995. *Research-Doctorate Programs in the United States,* edited by M. Goldberger, B. Maher, and P. Flattau. Washington, DC: National Academy Press.

National Research Council. 1998. *Trends in the Early Careers of Life Scientists.* Washington, DC: National Academy Press.

National Science Board. 2000. *Science and Engineering Indicators–2000.* Arlington, VA: National Science Foundation.

National Science Board. 2002. *Science and Engineering Indicators–2002.* Arlington, VA: National Science Foundation.

National Science Board. 2004. *Science and Engineering Indicators 2004.* Arlington, VA: National Science Foundation.

National Science Board. 2006. *Science and Engineering Indicators 2006.* Arlington, VA: National Science Foundation.

National Science Foundation. 1996. Undergraduate Origins of Recent (1991–95) Science and Engineering Doctorate Recipients, Detailed Statistical Tables. NSF 96–334. Arlington, VA: NSF Division of Science Resources Statistics.

National Science Foundation. Various years. National Center for Education Statistics Survey Databases. Arlington, VA: NSF Division of Science Resources Statistics. Data available on WebCASPAR. http://caspar.nsf.gov.

National Science Foundation. Various years. National Survey of College Graduates. Arlington, VA: NSF Division of Science Resources Statistics.

National Science Foundation. Various years (biennial). Survey of Doctorate Recipients. Arlington, VA: NSF Division of Science Resources Statistics.

National Science Foundation. Various years. Survey of Earned Doctorates. Arlington, VA: NSF Division of Science Resources Statistics. Data available on WebCASPAR. http://caspar.nsf.gov.

National Science Foundation. Various years. Survey of Graduate Students and Postdoctorates in Science and Engineering. Arlington, VA: NSF Division of Science Resources Statistics.

National Science Foundation. Various years. Survey of Research and Development Expenditures at Universities and Colleges. Arlington, VA: NSF Division of Science Resources Statistics.

Nelson, R. R., and N. Rosenberg. 1993. Technical Innovation and National Systems. In *National Innovation Systems,* ed. R. R. Nelson, 3–21. New York: Oxford University Press.

Newberg, J. A., and R. L. Dunn. 2002. Keeping Secrets in the Campus Lab: Law, Values, and Rules of Engagement for Industry-University R&D Partnerships. *American Business Law Journal* 39(2):187–240.

NIH News. 2003. NIH Calls on Scientists to Speed Public Release of Research Publications. http://www.nih.gov/news/pr/feb2005/od-03.htm (last accessed April 16, 2007).

NIH Web site. Award Data. http://grants1.nih.gov/grants/award/award.htm.

Normile, D. 2000. New Incentives Lure Chinese Talent Back Home. *Science* 287(5452):417–18.

————. 2001. Research Kicks into High Gear after a Long, Uphill Struggle. *Science* 291(5502):237–38.

North, D. 1995. Soothing the Establishment: The Impact of Foreign-Born Scientists and Engineers on America. Lanham, MD: University Press of America.

Owen-Smith, J. 2003. From Separate Systems to a Hybrid Order: Accumulative Advantage across Public and Private Science at Research One Universities. *Research Policy* 32(6):1081–1104.

Patent and Trademark Office. 2001. Utility Examination Guidelines. *66 Fed Reg. 1092* (January 5). http://www.uspto.gov/web/offices/pac/utility/utility.htm (last accessed April 16, 2007).

Payne, A. A. 2002. Do U.S. Congressional Earmarks Increase Research Output at Universities? *Science and Public Policy* 29(5):314–30.

Payne, A. A. 2006. Earmarks and EPSCoR: Shaping the Distribution, Quality, and Quantity of University Research. In *Shaping Science and Technology Policy: The Next Generation of Research,* ed. D. H. Guston and D. Sarewitz, 149–72. Madison: University of Wisconsin Press.

Pavitt, K. 2000. Why European Union Funding of Academic Research Should Be Increased: A Radical Proposal. *Science and Public Policy* 27(6):455–60.

Petersen, M. 2002. Madison Ave. Has Growing Role in the Business of Drug Research. *New York Times* (November 22): A1.

Petit, C. W. 1998. Germinating Access: Berkeley Department's Big Deal with Firm Aimed at Speeding Genetic Finds to Market. *U.S. News & World Report* (October 26): 60.

Phillips, M. 1996. Math Ph.D.s Add to Anti-foreigner Wave. *Wall Street Journal* (September 4): A2.

Plott, C. R. 1969. Some Organizational Influences on Urban Renewal Decisions. *American Economic Review* 58(2):306–21.

Pollack, A. 2006. Justices Reach Out to Consider Patent Case. *New York Times* (March 20): C1.

Pollan, M. 1998. Playing God in the Garden. *New York Times Magazine* (October 25): 45.

Porter, M. E., and D. Van Opstal. 2001. *U.S. Competitiveness 2001: Strengths, Vulnerabilities and Long-Term Priorities.* Washington, DC: Council on Competitiveness.

Press, E., and J. Washburn. 2000. The Kept University. *The Atlantic Monthly* 285(3):39–54.

Pressman, L., R. Burgess, R. M. Cook-Deegan, S. J. McCormack, I. Nami-Wolk, M. Soucy, and L. Walters. 2006. The Licensing of DNA Patents by U.S. Academic Institutions: An Empirical Survey. *Nature Biotechnology* 24(1):31–39.

Rahm, D. 1994. U.S. Universities and Technology Transfer: Perspectives of Academic Administrators and Researchers. *Industry and Higher Education* 8(2):72–78.

Rai, A. K. 1999. Regulating Scientific Research: Intellectual Property Rights and the Norms of Science. *Northwestern University Law Review* 94(1):77–152.

Rai, A. K., and R. S. Eisenberg. 2003. Bayh-Dole Reform and the Progress of Biomedicine. *American Scientist* 91(1):52–59.

Ramirez, Heather H. 2004. Defending the Privatization of Research Tools: An Examination of the "Tragedy of the Anticommons" in Biotechnology Research and Development. *Emory Law Journal* 53:359–89.

Rao, M. 1995. Foreign Students and Graduate Economic Education in the United States. *Journal of Economic Education* 26(3):274–81.

Ray, B. 1980. Congressional Promotion of District Interests: Does Power on the Hill Really Make a Difference? In *Political Benefits: Empirical Studies of American Public Programs*, ed. B. S. Rundquist, 1–35. Lexington, MA: Lexington Books.

Reiss, P., and D. H. Thurgood. 1991. *Summary Report 1001: Doctorate Recipients from United States Universities.* Washington, DC: National Academy Press.

Rizzo, M. J., and R. G. Ehrenberg. 2004. Resident and Nonresident Tuition and Enrollment Decisions at Flagship State Universities. In *College Choices: The Economics of Where to Go, When to Go, and How to Pay for It*, ed. C. Hoxby, 303–49. Chicago: University of Chicago Press.

Roche Prods, Inc. v. Bolar Pharmaceuticals Co. 1984. 733 F.2d 858 (Fed. Cir.), cert. denied, 469 U.S. 856 (1984).

Rosenberg, N. 2004. Science and Technology: Which Way Does Causality Run? Mimeo, Stanford University.

Rosenberg, N., and R. R. Nelson. 1994. American Universities and Technical Advance in Industry. *Research Policy* 23(3):323–48.

Rotemberg, J. J. 2002. Commercial Policy with Altruistic Voters. *Journal of Political Economy* 111(1):174–201.

Ruttan, V. W. 1982. *Agricultural Research Policy.* Minneapolis: University of Minnesota Press.

Sanford, J. C., E. D. Wolf, and N. K. Allen. 1990. Method for Transporting Substances into Living Cells and Tissue and Apparatus Therefor. U.S. Patent No. 4 945 050, July 31.

Savage, J. D. 1999. *Funding Science in America: Congress, Universities, and the Politics of the Academic Pork Barrel.* New York: Cambridge University Press.

Sears, R.G. 1998. Status of Public Wheat Breeding—1998. In *The Inaugural National Wheat Industry Research Forum*, ed. National Association of Wheat Growers et al., 80–82. Washington, DC: NAWG Foundation.

Shackelford, B. 2004. U.S. R&D Projected to have Grown Marginally in 2003. In *Info-Brief*, NSF 04-307, ed. Division of Science Resources Statistics. Arlington, VA: National Science Foundation.

Siegel, S. 1956. *Nonparametric Statistics for the Behavioral Science.* New York: McGraw Hill.

Silber, J. 1987. Testimony of Dr. John Silber, President of Boston University, before the Committee on Science, Space, and Technology, United States House of Representatives. Washington, DC (June 25).

Slaughter, S., and G. Rhoades. 1996. The Emergence of a Competitiveness Research and

Development Policy Coalition and the Commercialization of Academic Science and Technology. *Science, Technology and Human Values* 21(3):303–39.

Slaughter, S., and L. L. Leslie. 1997. *Academic Capitalism: Politics, Policies, and the Entrepreneurial University*. Baltimore, MD: Johns Hopkins University Press.

Smith, S. 1991. Shift Share Analysis of Change in Occupational Sex Composition. *Social Science Research* 20(4):437–53.

Stephan, P. E. 1996. The Economics of Science. *Journal of Economic Literature* 34(3):1199–235.

Stephan, P. E., G. Black, J. Adams, and S. G. Levin. 2002. Survey of Foreign Recipients of U.S. Ph.D.s. *Science* 295 (5563): 2211–12.

Stephan, P. E., S. Gurmu, A. J. Sumell, and G. Black. 2007. Who's Patenting in the University? Evidence from the Survey of Doctorates. *Economics of Innovation and New Technology* 16(2):71–99.

Stephan, P. E., and S. G. Levin. 1992. *Striking the Mother Lode in Science: The Importance of Age, Place, and Time*. New York: Oxford University Press.

———. 1996. Property Rights and Entrepreneurship in Science. *Small Business Economics* 8(3):177–88.

———. 2001. Exceptional Contributions to U.S. Science by the Foreign-Born and Foreign-Educated. *Population Research and Policy Review* 20(1–2):59–79.

———. 2003. Foreign Scholars in U.S. Science: Contributions and Costs. Prepared for the Cornell Higher Education Research Institute Science and the University conference (May 15–16), Cornell University, Ithaca, NY.

Stephan, P. E., and J. Ma. 2005. The Increased Frequency and Duration of the Postdoctorate Career Stage. *American Economic Review* 95(2):71–75.

Stephan, P. E., A. Sumell, G. Black, and J. Adams. 2004. Doctoral Education and Economic Development: The Flow of New PhDs to Industry. *Economic Development Quarterly* 18(2):151–67.

Stern, S. 2004. Do Scientists Pay to be Scientists? *Management Science* 50(6):835–53.

Stokes, D. 1997. *Pasteur's Quadrant*. Washington, DC: Brookings Institution Press.

Sung, L. M. 2000. Collegiality and Collaboration in the Age of Exclusivity. *DePaul Journal of Health Care Law* 3:411.

Symposium. 2002. The Human Genome Project, DNA Science and the Law: The American Legal System's Response to Breakthroughs in Genetic Science. Panel One (October 19, 2001): Intellectual Property and Genetic Science: The Legal Dilemmas. *American University Law Review* 51:371–99.

Tanksley, S. D., and S. R. McCouch. 1997. Seed Banks and Molecular Maps: Unlocking Genetic Potential from the Wild. *Science* 277(5329):1063–66.

Thursby, J. G., and M. C. Thursby. 2000. Today's Industry/University Licensing Environment: Issues and Problems as Reported by Participants. Paper presented at the Licensing Executive Society (U.S.A. & Canada), New York (May).

Thursby, J. G., R. Jensen, and M. C. Thursby. 2001. Objectives, Characteristics and

Outcomes of University Licensing: A Survey of Major U.S. Universities. *Journal of Technology Transfer* 26(1–2):59–72.

Thursby, J. G., and M. C. Thursby. 2002. Who is Selling the Ivory Tower? Sources of Growth in University Licensing. *Management Science* 48(1):90–104.

———. 2003. University Licensing and the Bayh-Dole Act. *Science* 301(5636):1052.

Thursby, M., J. Thursby, and S. Mukherjee. 2005. Are There Real Effects of Licensing on Academic Research? A Life Cycle View. National Bureau of Economic Research Working Paper 11497.

Trajtenberg, M. 1990 A Penny for Your Quotes: Patent Citations and the Value of Innovations. *Rand Journal of Economics* 21(1):172–87.

Trajtenberg, M., R. Henderson, and A. Jaffe. 1997. University versus Corporate Patents: A Window on the Basicness of Invention. *Economic Innovation and New Technology* 5(1):19–50.

Tripathi, M., S. Ansolabehere, and J. M. Snyder. 2002. Are PAC Contributions and Lobbying Linked? New Evidence from the 1995 Lobby Disclosure Act. *Business and Politics* 4(2), Article 2.

Vavilov, N. I. 1940. The New Systematics of Cultivated Plants. In *The New Systematics,* ed. J. H. Clarendon, 549–56. Oxford: Clarendon Press.

Washburn, J. 2005. University Inc.: The Corruption of Higher Education. New York: Basic Books.

Weiss, G. 2003. Age Bar Forces Europe's Senior Researchers to Head West. *Science* 302 (5652):1885–86.

Winston, G. 1999. Subsidies, Hierarchies and Peers: The Awkward Economics of Higher Education. *Journal of Economic Perspectives* 13(1):13–36.

Wright, J. R. 1996. Interest Groups and Congress: Lobbying, Contributions, and Influence. Boston, MA: Allyn and Bacon.

Yamagata, H. 1999. Trends in Faculty Attrition at U.S. Medical Schools, 1980–1999. *Analysis in Brief* 2(2):1–2.

Yimin, D. 2001. In China, Publish or Perish Is Becoming the New Reality. *Science* 291(5508): 1471–79.

Ziman, J. M. 1994. *Prometheus Bound: Science in a Dynamic Steady State.* New York: Cambridge University Press.

Zucker, L. G., M. R. Darby, and J. Armstrong. 1998a. Geographically Localized Knowledge: Spillovers or Markets. *Economic Inquiry* 36(1):65–86.

Zucker, L. G., M. R. Darby, and M. B. Brewer. 1998b. Intellectual Human Capital and the Birth of U.S. Biotechnology Enterprises. *American Economic Review* 88(1):290–306.

Zucker, L. G., M. R. Darby, and J. Armstrong. 1999. Intellectual Capital and the Firm: The Technology of Geographically Localized Knowledge Spillovers. National Bureau of Economic Research Working Paper 4946.

Zumeta, W., and J. S. Raveling. 2003. Attracting the Best and the Brightest. *Issues in Science and Technology* 19:36–40.

Contributors

GRANT C. BLACK is assistant professor of economics and director of the Bureau of Business and Economic Research and Center for Economic Education at Indiana University South Bend.

GEORGE J. BORJAS is the Robert W. Scrivner Professor of Economics and Social Policy at the Kennedy School of Government, Harvard University, and a research associate at the National Bureau of Economic Research.

W. RONNIE COFFMAN is the International Professor of Plant Breeding and Genetics and director of international programs in the College of Agriculture and Life Sciences at Cornell University.

JOHN M. DE FIGUEIREDO is associate professor of policy at the Anderson School of Management at the University of California, Los Angeles, and a faculty research fellow at the National Bureau of Economic Research.

RONALD G. EHRENBERG is the Irving M. Ives Professor of Industrial and Labor Relations and Economics and director of the Cornell Higher Education Research Institute at Cornell University and editor of *What's Happening to Public Higher Education.*

RICHARD B. FREEMAN is the Ascherman Professor of Economics at Harvard University, director of labor studies at the National Bureau of Economic Research, and a professorial research fellow at the Centre for Economic Performance, London School of Economics.

HOWARD GARRISON is director of the Office of Public Affairs of the Federation of American Societies for Experimental Biology.

SUSAN A. GERBI is the George Eggleston Professor of Biochemistry, Department of Molecular Biology, Cell Biology and Biochemistry, Brown University.

JEFFREY A. GROEN is a research economist at the Bureau of Labor Statistics, U.S. Department of Labor, and a faculty associate at the Cornell Higher Education Research Institute.

DIANA M. HICKS is professor and chair of the School of Public Policy, Georgia Institute of Technology.

GEORGE H. JAKUBSON is associate professor of labor economics at Cornell University and faculty associate at the Cornell Higher Education Research Institute.

EMILY JIN is an associate consultant at ZS Associates and a former research assistant at the National Bureau of Economic Research.

WILLIAM H. LESSER is the Susan Eckert Lynch Chair in Science and Business and Chair, Department of Applied Economics and Management, Cornell University.

SHARON G. LEVIN is professor emeritus of economics at the University of Missouri–St. Louis.

RISA L. LIEBERWITZ is associate professor of labor and employment law at the Cornell University School of Industrial and Labor Relations and a co-director of the Cornell University Law and Society concentration.

SUSAN R. MCCOUCH is professor of plant breeding and genetics at Cornell University and a faculty associate of the International Agriculture Program at Cornell.

MICHAEL J. RIZZO is a senior economist and director of summer programs at the American Institute for Economic Research in Great Barrington, Massachusetts.

CHIA-YU SHEN is a research assistant at the National Bureau of Economic Research Science and Engineering Workforce Project.

BRIAN S. SILVERMAN is the J.R.S. Prichard and Ann Wilson Chair in Management, Rotman School of Management, University of Toronto.

PAULA E. STEPHAN is professor of economics at Georgia State University and coeditor of the two-volume *Economics of Science and Innovation.*

JERRY G. THURSBY is the Goodrich C. White Professor of Economics, chair of the Department of Economics, and director of graduate studies at Emory University.

MARIE C. THURSBY is the Hal and John Smith Chair of Entrepreneurship in the College of Management, Georgia Institute of Technology, and a research associate at the National Bureau of Economic Research.

Index

academic freedom: and peer review, 56; and the public interest, 56; threats from corporate influence, 63

academic research: and academic patenting, 230–31, 239; and bibliometrics, 227, 229, 242n1; and changes in science at universities, 55–56; characteristics of, 56; and corporate influence, 62; critiques of commercialization of, 61–64; and databases, 227; and decline in U.S. output, 223–42; and demography, 231; and electronic publishing, 228–30; and ethical issues, 241–42; and explanations for U.S. decline, 226–28, 240–42; and federal funding, 232, 239; and goals of the university, 56; and hybrid paper-electronic (p-e) journals, 228; and impact of Bayh-Dole Act, 61; impact of licensing, 77–79; importance of, 6; and international graduate programs, 234–35; and knowledge-based economies, 233, 240–41; and medical publishing, 232; and metrics of performance, 6; and number of U.S. graduate students, 237–38; and personal advantages of publication, 241; and the public domain, 69–72; for the public good, 72; and public trust, 62; and publication in peer-reviewed journals, 223, 227–28; and publication counts, 223–24, 233; and publication output by field, 225; and publication output by sector, 225; and publication prizes in China, 236; and publishing worldwide, 232–34, 239; and pure electronic journals, 228; and quality factor in science, 227; and quantity of U.S. papers, 224, 239; and R&D expenditures, 237–38; and research funding, 231–35; and S&E growth, 223–24; and salaries, 236

Adams, J. D., 7, 79

advanced materials, advances in, ix, 20

aerospace engineering: and Ph.D. students at top-ten programs, 117–18, and stay rates, 127

African Americans. *See* blacks

age: and disclosure, 87–90; of Ph.D.s, 198; of scientific workforce, 156, 160; and stay rate, 120, 127–28

Agrawal, A., 79, 81, 230

Agricultural Research Service (ARS): focus on basic science, 95; and plant breeding R&D, 95

Science and Technology in Society

Scott Frickel and Kelly Moore, editors
The New Political Sociology of Science: Institutions, Networks, and Power

David H. Guston and Daniel Sarewitz, editors
Shaping Science and Technology Policy: The Next Generation of Research

Daniel Lee Kleinman
Impure Cultures: University Biology and the World of Commerce

Daniel Lee Kleinman, Abby J. Kinchy, and Jo Handelsman, editors
Controversies in Science and Technology: From Maize to Menopause

Jack Ralph Kloppenburg Jr.
First the Seed: The Political Economy of Plant Biotechnology, second edition

Paula E. Stephan and Ronald G. Ehrenberg, editors
Science and the University